How Humanity Turned Electricity into Electronics

From Rubbing Amber to Swiping Glass

Clemens Valens

This is an Elektor Publication. Elektor is the media brand of
Elektor International Media B.V.
PO Box 11, NL-6114-ZG Susteren, The Netherlands
Phone: +31 46 4389444

Declaration

ISBN 978-3-89576-664-0 Print
ISBN 978-3-89576-665-7 eBook

www.elektor.com
Editor: Clemens Valens
Prepress Production: D-Vision, Julian van den Berg, Harmen Heida, Elektor
Cover illustration by: Dimitri Jansma, www.studioparaat.nl
Printers: mediaprint solutions GmbH, Eggertstrasse 28, 33100 Paderborn, Germany

Table of Contents

Introduction

Electricity has been known to humanity for many centuries, even though the phenomenon wasn't understood or even named. It wasn't until the 17th century that scientists began to look into the curious effects caused by what we now call electricity. Progress was made quickly during the two centuries that followed and by the end of the 19th century, electricity entered the lives and homes of the 'common' people. In the 20th century, discoveries and inventions continued and the electronics industry developed at an ever-increasing pace.

As of 2025, tiny computer chips integrating millions of transistors, even billions for some, allow people all over the world and beyond to communicate and share information instantaneously, 24/7. The history of electronics is relatively short—250 years, say. If this exponential progress continues, who knows what future generations may come up with? I won't even try to guess, let alone predict.

While reading this book, you will come across a great number of individuals with

From electricity to electronics. The power of lightning captured in a Leyden jar by Benjamin Franklin in 1752 gave birth, 250 years later, to powerful microcontrollers in tiny packages running applications based on artificial intelligence.

no shortage of brains and curiosity. These were the sorts who, when faced with something unusual, wouldn't rest until they had made sense of it, whether by discovering a new phenomenon, laying down a law, or dreaming up some ingenious device. Some have even been hailed as the greatest minds of their field, their era, or even all of history. That might be stretching things a bit, but it's fair to say they were a sharp lot with a strong interest in the unknown. Indeed, whenever something peculiar popped up—be it an apple plummeting earthward or a kettle letting off steam—you could be sure one of these chaps would turn up, finger raised, ready to exclaim, "Aha! What have we here?"

While reading this book, you may be surprised to find that some of these brilliant minds are mentioned with full credentials, while others are notably absent. This is simply the result of storytelling. Only the people that fit in the story are named.

A startling observation I made while reviewing the manuscript of this book was that it mentioned only two women—and only in passing, as they didn't contribute to electronics. This does not mean, of course, that no women have played an important role in electronics—only that I have not encountered any that fit inside the scope of this book. Let us, however humbly, attempt to redress this by reminding that each man has a mother who contributed half of his genetic makeup.

An index of all the individuals referenced in this book is provided at the end.

Girl Power

In 1941, together with American composer, pianist and inventor George Johann Carl Antheil (1900–1959), the Austrian actress and inventor Hedwig Eva Maria Kiesler (1914–2000), better known as Hedy Lamarr, filed a patent for a secret communication system which featured frequency-hopping spread-spectrum technology (FHSS). Today, this kind of technology is commonly used in wireless communication systems.

Just So You Know

In this book, I do not aim to explain how electronic components and circuits work. There are no exercises to complete, nor will you find difficult equations to solve. Reading this book will not teach you a new skill. The goal of this book is simply to entertain the reader with, hopefully, interesting—though not always useful—background information about the history of electronics and some of the themes and concepts considered significant in the field. While a little bit of mathematics is included in some chapters, it can easily be skipped without missing anything essential. This book does not aspire to be complete or exhaustive (though perhaps exhausting at times), and it is not a history book either. All the subjects covered in this book have been thoroughly researched, and I believe the material presented in the following pages is accurate. However, there may still be mistakes. If you find one, please don't feel compelled to let me know.

This book draws on common knowledge and cultural references, as found in popular belief and sources like Wikipedia and encyclopaedias. However, this does not guarantee that everything is true. It is well known that many discoveries and inventions have been made independently by multiple individuals in different locations, sometimes separated by decades. This is often due to a lack of communication—some scientists or inventors did not (or do not) publish their results, while others failed to publish in the appropriate channels. Consequently, many events are misattributed, and disputes over patents and inventions have been common. While some historians have attempted to correct these inaccuracies, it is difficult to do so retroactively. Moreover, even the sources consulted may themselves be biased, and verifying them can be a complex and time-consuming task.

As you may notice, this book has a strong Western bias, which is due to the sources available to me. I have consulted numerous patent applications and original publications whenever they were accessible (and legible), but I have not travelled the world to visit libraries or interview individuals—partly because most of the people mentioned in this book passed away many years ago. As a result, I may have missed obscure references here and there. For all of the above reasons, I kindly ask the reader for their indulgence.

The Electron

This book is about electronics. But what exactly is electronics and how did it come about? To find out, a good place to start is at the beginning, by its definition or, better, definitions, as the English word 'electronics' has several. First of all, electronics is the branch of science or technology that deals with the study of the flow of electrons through various mediums. Electronics is also defined as the use of such technology in order to produce equipment that uses electricity in order to work. Finally, the word electronics is used for the circuits or parts of equipment that use electricity, like transistors, chips, etc.

As is clear from the word 'electronics', the electron sits at its base. Without the electron, the electrical current as we know it would not exist. Without electrical current, electric and electronic equipment doesn't work very well, and this is why you pay the bill of the electricity company every month. Yet the use of electricity and the development of devices and machines using it started long before the discovery of the electron. Until then, the atom was thought to be the basic building block of matter.

The history of electricity goes back thousands of years. Even though some people claim that ancient astronauts and unknown civilizations that no longer exist made use of electricity, one of the first recorded observations related to electricity was made by the Greek philosopher Thales of Miletus, somewhere around 600 BC. He knew that amber, when rubbed, can attract tiny lightweight objects. Amber, fossilized tree resin, was called 'electron'[1] in ancient Greek (ἤλεκτρον, excuse my pronunciation).

The electron: may it never be of any use to anybody!

J. J. Thomson

The concept of the atom dates back to ancient Greek philosophers like Democritus and Leucippus who proposed in the 5th century BC that all matter is made up of tiny, indivisible particles that they called 'atoms' (from the Greek word 'atomos,' meaning 'uncuttable'). This purely philosophical idea got experimental support only in 1803 when John Dalton published his atomic theory.

John Dalton (1766–1844) was an English chemist, physicist, and meteorologist who conducted experiments with gases under various conditions. He observed that, in chemical reactions, the same elements always combined in the same way depending on their masses. This suggested that elements consisted of discrete units (at-

1 The word electron is a relative of another ancient Greek word 'elektor' (ἠλέκτωρ), which means 'bright', 'radiant', 'shining' when referring to the Sun.

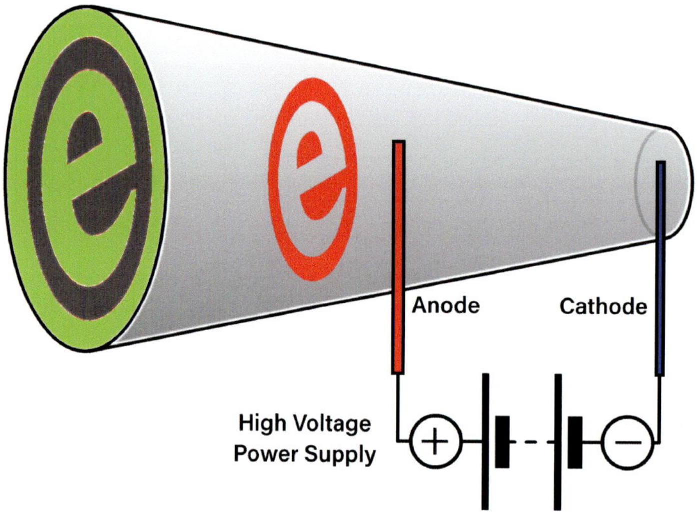

An object (the Elektor logo) placed inside a Crookes tube casts a shadow on the anode (left) side of the tube.

oms) that combined in fixed proportions. Pushing things a bit further, he found a way to compare the masses of different atoms. Based on his observations, Dalton suggested that each element consists of identical atoms and that atoms combine in specific ratios to form compounds. The smallest atom is hydrogen.

This was the state of affairs until British physicist Joseph John (J.J.) Thomson (1856–1940) started his work on cathode rays. These mysterious rays were discovered a few years before when researchers began to experiment with high voltages in a vacuum. In 1857, German physicist and glassblower Heinrich Geissler (1814–1879) had created a glass tube with a low vacuum and two electrodes in it, an anode and a cathode. When a high voltage was applied between the electrodes, the Geissler tube started to glow. English chemist and physicist William Crookes (1832–1919) perfected the tube and the vacuum in it. This resulted in the Crookes tube. This tube only glows on the anode end when a high voltage is applied to it. Objects inside the tube, like the anode, casted shadows, indicating that the glow was caused by particles flying inside the tube.

An object (the Elektor logo) placed inside a Crookes tube casts a shadow on the anode (left) side of the tube.

Two theories were developed to explain the effect. To William Crookes, the effect was due to 'radiant matter', consisting of electrically charged atoms, whereas German physicists Heinrich Rudolph Hertz (1857–1894) and Eugen Goldstein (1850–1930) proposed a new form of electromagnetic waves called 'aether vibrations'. J.J. Thomson's goal was to try to settle this controversy.

Using an improved vacuum cathode ray tube, Thomson managed to show that the glow effect was indeed caused by charged particles, but by particles lighter than hydrogen by more than a thousand times. Furthermore, he found that the effect occurred in the same way for different gases and metals used in the tube. This led him to believe that a particle common to all matter was responsible for the effect. When he found that these particles could also be produced in other ways, by heating metal for instance, he was convinced. Thomson called this particle 'corpuscle' but the scientific community preferred 'electron', a name that had been proposed in 1891 by the Irish physicist George Johnstone Stoney (1826–1911) to describe the fundamental unit of electrical charge. The official year of discovery of the electron is 1897.

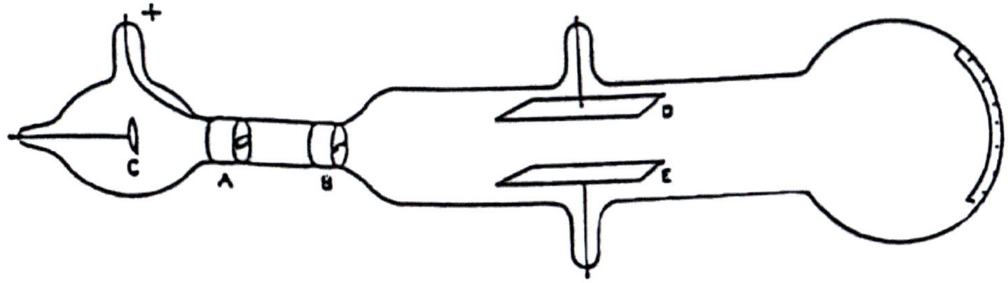

A drawing by J. J. Thomson of the vacuum tube he used to discover the electron in 1897.

Capacitance & the Leyden Jar

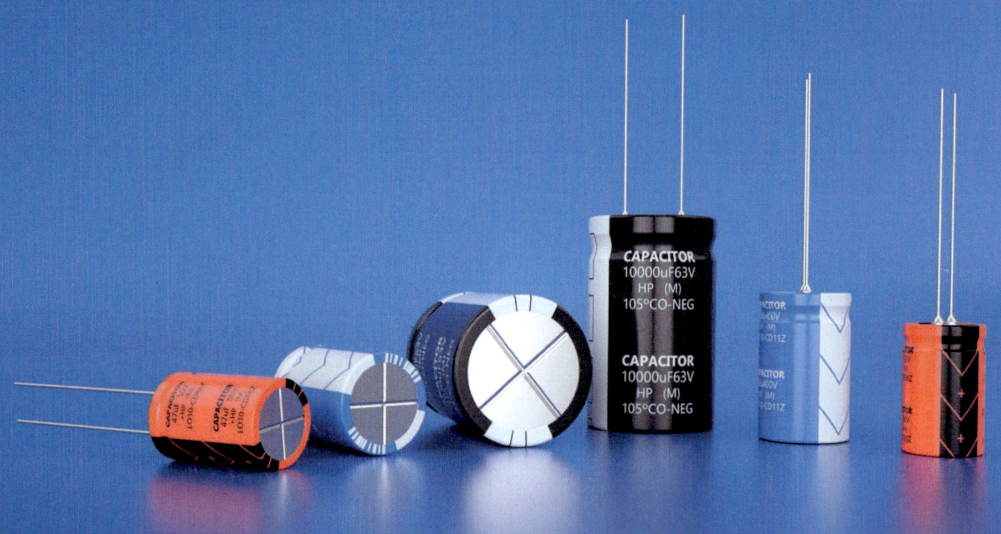

The observation that rubbing amber and other materials created an unknown force, sparked a lot of interest in the scientific world of the 17th century. First, one tried to improve the techniques to create this force with so-called friction machines that were based on the rubbing principle. The first of these machines is attributed to German scientist, inventor, mathematician and physicist Otto von Guericke (1602–1686), who used a sulphur globe that could be rotated and rubbed by hand. Isaac Newton (1642–1726[2]) suggested the use of a glass globe instead of a sulphur one. When friction machines became better, the scientific goal shifted from generating the static electricity to storing it. German experimenter Georg Matthias Bose (1710–1761) is considered to be the first to have successfully stored static charge temporarily. He introduced an insulated conductor (called a prime conductor), which consisted of an assistant standing on a block of resin (an effective insulator) holding a metal bar in one hand while touching the spinning globe with the other.

The Flying Boy Experiment

Friction machines became popular attractions at science fairs and other events in the 18th century where they were used to demonstrate static electricity in amusing and spectacular ways. English dyer and astronomer Stephen Gray (1666–1736), who showed that electricity can be conducted through metals and that it appeared on the surfaces of insulators, imagined the 'Flying Boy' experiment. In this experiment, a boy suspended from silk ropes was given an electrical charge and then attracted pieces of paper and chaff with his hands. People from the audience were encouraged to touch the boy's nose to experience an electric shock. A hugely popular variant of the Flying Boy demonstration was the 'Electric Kiss' by Georg Matthias Bose. Here, the suspended boy was replaced by an electrically charged young woman standing on an isolated platform. Then a young man from the audience was invited to give her a kiss, which resulted in the pair receiving an electric shock.

The Flying Boy experiment reproduced by Jean-Antoine Nollet a.k.a. Abbé Nollet (1700–1770), the inventor of the electroscope.

2 The exact dates of Newton's birth and death are complicated as he lived in a period when two calendars were in use in Europe: Julian and Gregorian.

If anyone should doubt whether the electrical matter passes through the substance of bodies, or only over along their surfaces, a shock from an electrified large glass jar, taken through his own body, will probably convince him.

Benjamin Franklin

Inspired by Bose's work, German jurist, Lutheran cleric, and physicist Ewald Georg von Kleist (1700–1748) was convinced that he could capture static electricity in a glass bottle. The glass would provide an isolation barrier for the electricity. He filled the bottle with some water to store the electricity in. The bottle was kept in one hand while 'pouring' the electricity generated with a friction machine into the bottle through a needle stuck through its cork. When, during one of his experiments, he accidentally touched the needle, the electric shock he received proved his assumptions.

The results obtained by Von Kleist were diffused in a somewhat confusing way to the scientific community. They were picked up in, amongst other places, Leiden, in the Netherlands, by Dutch scientist Pieter van Musschenbroek (1692–1761). He started experimenting and obtained similar results as Von Kleist. Besides being a scientist, Pieter van Musschenbroek also manufactured and sold scientific equipment, and he added the bottle to his catalogue. His distributor in France, Jean-Antoine Nollet (1700–1770) better known as Abbé Nollet, started selling the electrical storage device as the Leyden jar, which is how it got its name. It is also known as the Kleistian jar.

American scientist (and much more than that) Benjamin Franklin (1706–1790) experimented with Leyden jars connected in series. He referred to his contraptions as electrical batteries. Today, the term 'battery' is used for combinations of multiple electrochemical cells.

Nollet conducted an interesting experiment involving about two hundred monks and a battery of Leyden jars. The monks were placed in a circle with pieces of iron wire connecting them. Nollet then discharged the battery through the human chain. When all the monks sprang into the air almost simultaneously, Nollet concluded that the speed of propagation of the electricity was very high.

The Leyden jar became widely used as a means of storing static electricity and marked a significant step forward in the study of electricity. These early experiments with the Leyden jar laid the groundwork for the capacitor, one of the three passive components that are at the foundations of electronics. The other two are the resistor and the inductor.

In the last decade, the use of capacitance has become popular in human user interfaces like touch screens on tablets and smartphones, and as non-contact buttons

Make Your Own Leyden Jar

With a few common household items, you can make your own Leyden jar. Several approaches are possible, but an easy one that worked well for me is as follows. Take a small, empty plastic water bottle that still has its plastic cap. The bottle should not have patterns or embossments like ridges and dents, the smoother the surface the better. Wrap the bottle in aluminium foil while keeping its neck free. Put a tiny bit of kitchen salt in the bottle and fill it with tap water to about 3 to 4 cm under the lip. Screw the cap on and shake the bottle to mix the water and the salt. Push a long nail or screw through the centre of the cap down into the water. It should be a few centimetres deep in the water Now it is time to charge the jar. A length of PVC pipe rubbed with a woollen cloth works well. You can also try with paper towel. Let an assistant hold the bottle in one hand. Rub the pipe vigorously and then move it closely along the head of the nail without touching it. If you listen carefully, you will hear static clicks when the charge transfers into the bottle. Repeat this a few times. Five to ten times should be enough. Now, slowly move your finger close to the head of the nail. As you come really close, a spark may jump to your finger or, if you touch the nail, you will feel an electric shock.

on many household appliances. These interfaces use the fact that the human body is composed of water and electrolytes (salts), which make it a good electrical conductor. The touch-sensitive surface behaves as one plate of a capacitor. A finger close to it acts

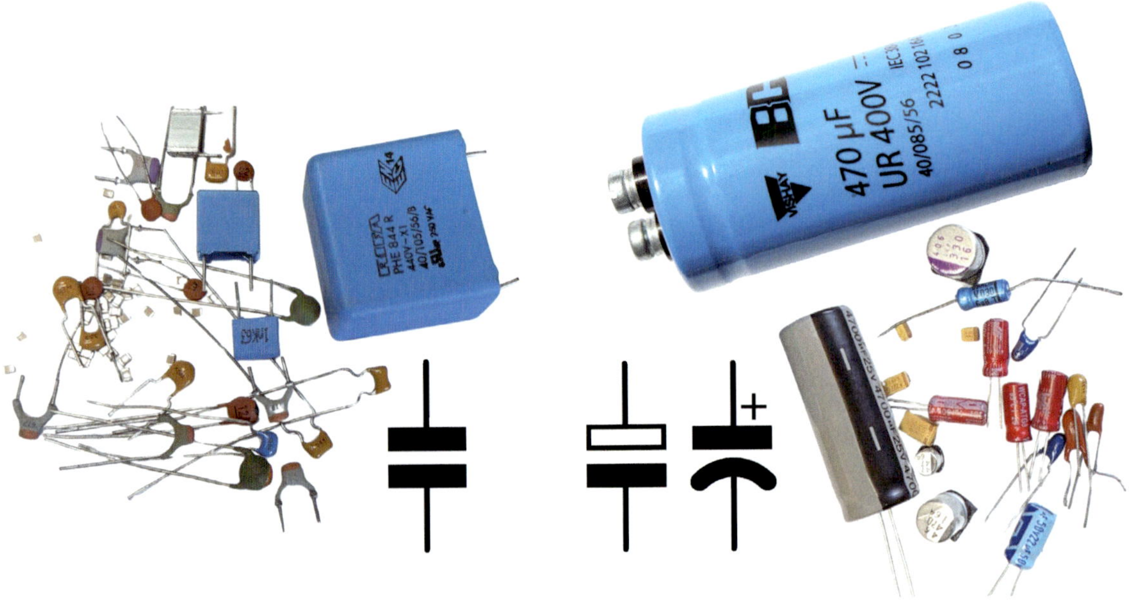

The electric symbol of the capacitor as it is used today in circuit diagrams and schematic drawings together with a collection of capacitors. Non-polarized capacitors on the left, polarized capacitors on the right. The polarized symbols come in two flavours: European (left, obsolete according to Wikipedia) and US (right).

as the capacitor's second plate. The circuit connected to this capacitor can detect the change in capacitance and respond to it.

Capacitance (C) is measured in farad (F), named after Michael Faraday. To understand the use of capacitors in electronic circuits, it is useful to look briefly at the maths describing the device. The current-voltage relationship of a capacitor is given by:

$$I(t) = C\frac{dV(t)}{dt}$$

Here, I is the current through the capacitor (in ampere), t is the time (in seconds), C is the capacitance of the device (in farad), and V is the voltage over the capacitor (in volt). This equation states that the current through the capacitor is proportional to the rate of change of the voltage over it (indicated by the dV(t)/dt part). The equation may be more meaningful after integrating both sides of it:

$$\frac{1}{C}\int I(t)dt = V(t)$$

This shows that a capacitor can be charged by passing a current through it. The time to charge the device is inversely proportional to the value of the capacitor. In other words, the larger the capacitor, the longer it takes to charge it to a certain voltage level. It is important to note that these equations are time dependent. Because time and frequency are two aspects of the same phenomenon, this implies that a capacitor in a circuit introduces frequency-dependent behaviour. In many cases, this is exactly what we want, but not always. A filter, for example, relies often on capacitors to attenuate or amplify a range of frequencies contained in a signal. At the same time, parasitic capacitances introduced by wiring and the non-ideal components that make up an amplifier introduce unwanted behaviour at certain frequencies. This may for instance limit the usable frequency range or even produce instabilities.

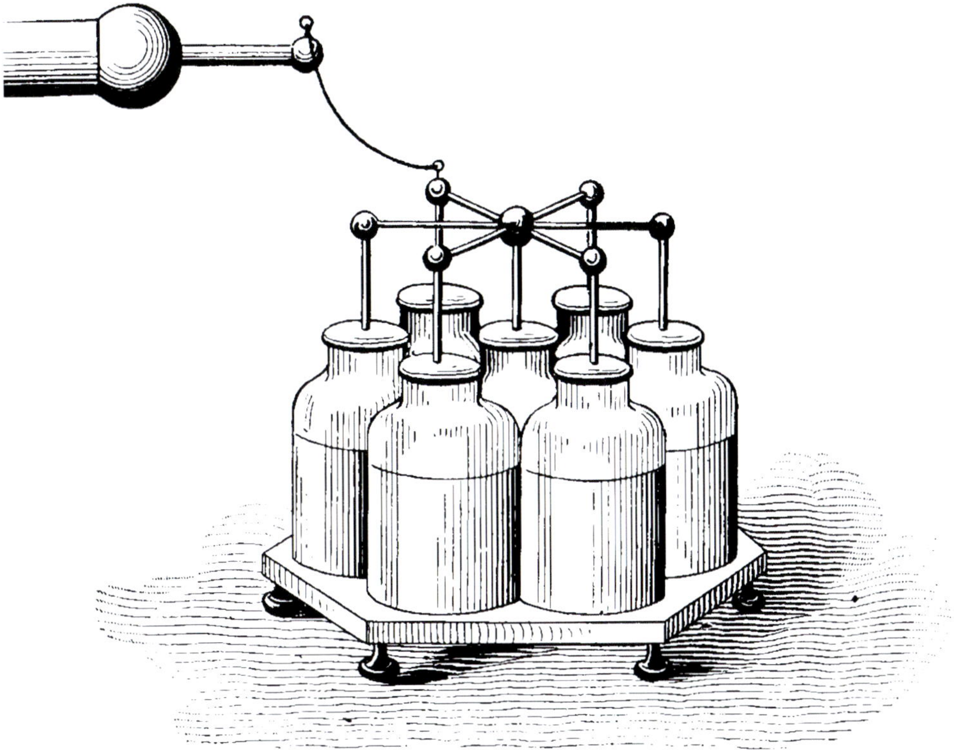

Invented by Benjamin Franklin, the Leyden battery allows storing more charge than a single Leyden jar."

The Battery & the Voltaic Pile

The electric battery, that practical device that led to the invention of the pocket torch and which is used today in its rechargeable shape to start and power cars and telephones with, has been at the origin of controversies ever since its invention. It all started with Italian physicist and chemist Alessandro Giuseppe Antonio Anastasio Volta (1745–1827) not agreeing with his contemporary fellow scientist Luigi Galvani (1737–1798) when the latter tried to explain the results of his famous frog-leg experiments.

Galvanic phenomena, named after Luigi Galvani, had already been observed in the 18th century, years before Galvani entered the arena. These observations and experiments involved Leyden jars, lightning, different metals, and living tissue like the human tongue and other muscles, but the link with electricity was unknown. When, in 1780, Galvani discovered that he could make a frog's leg contract by touching the iron plate the frog laid on with brass electrodes attached to the frog's spinal cord, he was convinced that he had found the electric power that animates living things. He called the effect 'animal electricity'.

Alessandro Volta did not agree. To him, the electricity was not produced inside the biological tissue, but instead by the contact between the two different metals. According to Volta, the animal cells were mere conductors. To prove this, he repeated Galvani's experiments with the frog's leg replaced by a piece of paper soaked in salt water (the electrolyte) and observed electricity flowing through the setup. Pushing things a bit further, he placed several of these cells in series. In 1800, simplifying even more, he stacked discs of dissimilar metals on top of each other, and separated them with pieces of cardboard soaked in salt water. The Voltaic pile, today considered to be the first electrochemical battery, was born.

Galvani's work greatly influenced Mary Shelley when she wrote Frankenstein in 1818.

From this point on batteries started to evolve quickly. All sorts of combinations of metals were tried with different kinds of electrolytes, and with more or less promising results. Two of these results are worth mentioning here. First of all, the replacement of the liquid electrolyte by a solid type. The dry cell, as opposed to the then common wet cell, was invented in 1886 by the German scientist Carl Gassner (1855–1942) . This was an important step forward as it made the battery portable and usable in any position and orientation. The concept was improved by E. M. Jewett, employee of the National Carbon Company (NCC), an American manufacturer of batteries and lighting products. As a result, NCC began

Make Your Own Voltaic Pile

The voltaic pile can be easily made at home. All you need are a few pieces of two different metals, bits of cardboard or paper towel, vinegar and kitchen salt. A piece of aluminium foil will be helpful too. For my voltaic pile, I used six bronze coins and six galvanized washers because that is what I had readily available.

Pour some vinegar in a cup or saucer and add the salt. Stir to dissolve as much of the salt as possible. Cut the cardboard into pieces roughly the size of the coins. Precision is not important. Lay the piece of aluminium foil on a flat, horizontal surface. This will be one of the electrodes, so it should stick out a bit. Place a coin on the aluminium foil. Soak a piece of cardboard in the vinegar-salt mixture and place it on top of the coin. Then place the washer (or other metal) on the piece of soaked cardboard. Continue stacking layers of coin-cardboard-washer (in that order) until all the coins have been used. The top should be a washer. With a multimeter, you should now be able to measure a potential difference of several volts, depending on the number of layers in the stack. I measured almost three volts between the aluminium foil and the top washer. The voltage dropped quickly, probably because instead of cardboard I had used very thin paper towel that sagged through the holes of the washers. Also, I hadn't put much effort into building my pile. Yet, it did work.

commercializing a dry, zinc-carbon, cylindrical 1.5 V battery. In 1898, the company launched the first D-type battery, a type that is still in use today (standardized as LR20) and that led in 1899 to the invention of the torch by English engineer David Misell (1846–1920). As a sidenote, David Misell was a prolific and creative inventor who filed many patents for electric devices like the electrically illuminated clock (1894, a kind of backlight from the front), a burglar alarm (1887, patented as 'electric battery'), an electric cigar lighter (1894), and a sort of electric candle (1895).

The second major improvement of the electrochemical battery was the invention of the rechargeable battery. This achievement is attributed to the French physicist Gaston Planté (1834–1889) who invented the lead-acid battery. Until then, batteries would be permanently drained when all their chemical reactants were spent. Planté's battery could be recharged by passing a current through it instead of draining a current. In 1881, French chemical engineer Camille Alphonse Faure (1840–1898) significantly improved Planté's battery and turned it into a commercial success. The electric car was born. Ironically, today the lead-acid battery is mainly used in non-electric vehicles.

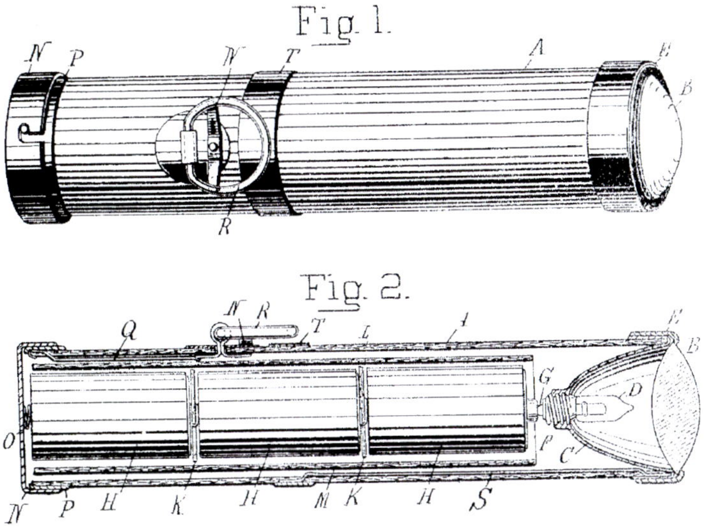

The D-type battery enabled the invention of the torch.

Now, this is all very interesting and such but let us go back to the controversies that we mentioned previously. As said before, Alessandro Volta did not agree with Galvani, but Galvani did not agree with Volta either. This gave rise to the infamous Galvani-Volta controversy that would last for some 150 years. Many papers and books have been written about it and it is has become a standard subject for science philosophers to study.

The controversy started with Galvani and Volta, but they didn't play an important role in it. The object of the controversy is the mechanism that produces the electricity in the electrochemical battery. For Volta it was clear that the electricity was generated due to two different metals making contact. The electrolyte was just a con-

Each metal has a certain power, which is different from metal to metal, of setting the electric fluid in motion.

Alessandro Volta

ductor providing the contact between the two metals. Here we will refer to it as the Contact Theory. Galvani on the other hand believed that the electricity was generated inside the electrolyte and the two metals allowed to get to it. We will call it the Chemical Theory.

Interesting in this controversy is that renowned scientists took opposing standpoints. On the contact-theory side, we find, besides Volta, names as Ohm, Davy, Gilbert, and Thomson (a.k.a. Lord Kelvin) whereas other illustrious scientists like Faraday, Becquerel and Ørsted supported Galvani's side. Many scientists elaborated 'crucial' experiments to prove their point of view forever, but in the end none of these experiments did. One of the main arguments against the contact theory was that it didn't respect the law of energy conservation. Simply by placing two different metals against each other, an inexhaustible current[3] would start to flow. The most important argument against the chemical theory was the fact that the contact approach didn't need chemicals for a current to flow. It took about 150 years to develop a theory that could explain the workings of an electric battery in a satisfactory way, simply because the required instruments to measure everything properly didn't exist before. Now we know that both theories were partly right. Simply put, when two different metals (the electrodes) are brought into contact, directly or through a conductor (an electrolyte), a voltage difference develops between the two electrodes. This voltage attracts charge present in the electrolyte and free electrons accumulate at the positive electrode (the anode). When a conductor is connected between the two electrodes, the electrons will flow from the anode to the other electrode, the cathode, where they recombine. When all the charge has been consumed, the electrolyte has become an insulator, and the battery is empty.

Other controversies due to the electric battery concern environmental and geopolitical issues. Used batteries left to rot out in the open cause chemical pollution, therefore they must be disposed of in a good way. Manufacturing batteries requires rare metals and minerals but mining these causes slavery, child labour, wars and pollution. Yet the number of battery-powered machines, devices and gadgets continues to increase. Electric vehicles needing powerful batteries are supposed to replace combustion motors while thousands, if not millions of battery-powered Internet of

3 At the start of the Galvani-Volta controversy, electric current hadn't been discovered yet and was referred to as 'electric fluid'.

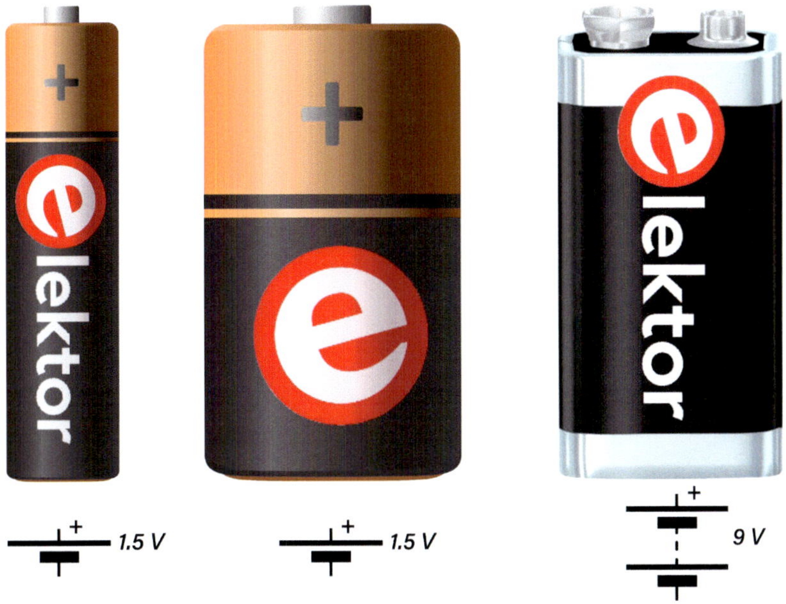

Three common batteries with the electric symbol. From left to right the 1.5 V AA or LR6, the 1.5 V C or LR14 and the 9 V PP3 type. Inside the PP3 are usually six small 1.5 V cells connected in series, which is why its electric symbol consists of more than one cell.

The Law of Capacitance Gave Us Volts

Besides starting the Galvani-Volta controversy, Alessandro Volta laid the foundation for electrochemistry and the development of batteries. Luigi Galvani's work on the other hand inspired further research into bioelectricity, leading to the discovery of electrical signals in nerves and muscles, one of the foundations of neurobiology.

Volta also explored what we now refer to as electrical capacitance. He discovered that for a given object the electrical potential difference (*V*) and charge (*Q*) are proportional. This is called Volta's law of capacitance. It is said that the unit of electrical potential and of voltage (the difference between two electrical potentials) has been named the volt because of this work. What happened to the 'a' in Volta is not known.

Things (IoT) edge nodes (sensors) are being installed each day. And what about smartphones? It will probably take a few decades or more to resolve these controversies too.

The battery is a great invention, and it allowed for the creation of all sorts of cool gadgets, but the fact that it doesn't last forever makes it unpractical for many applica-

This silver coin commemorating the war of the currents. It is the first in a series of four dedicated to the life and work of Nikola Tesla, issued in 2023 by the Czech Mint.

tions. Powering a home for instance is still best done with a power generator installed at an electricity plant. This is where the famous names of Thomas Alva Edison (1847–1931), George Westinghouse Jr. (1846–1914), and Nikola Tesla (1856–1943) enter our story.

Thomas Edison, probably best known for the incandescent lightbulb he did not invent, was in favour of a direct current (DC) power grid. According to him, a 100-volt DC supply in every house would be safe (enough). He added 10% to it to accommodate for wire losses, which is why today many countries have a 110-volt-based mains supply system (albeit AC). He founded the Edison Electric Light Company to develop and install DC systems for electric lighting and power distribution.

George Westinghouse, after having experimented with DC and being hindered by Edison's DC patents, was convinced that alternating current (AC) had more potential as an electrical system. He founded the Westinghouse Electric Corporation and started to install AC-lighting systems and power generators all over the USA. Westinghouse teamed up with Nikola Tesla who had invented the AC-powered induction motor and the polyphase AC system. The War of the Currents had begun, yet another controversy related to electrical power.

The war of the currents raged for several years. One of the main battles involved the electric chair that, at the time, was being investigated as a more humane means of execution. One question that hadn't been answered yet was what the best power source would be for it. Edison's main argument against AC was that it was more dangerous than DC, and so he proposed AC power. He even recommended Westinghouse as the energy supplier. Westinghouse didn't want to be associated with the electric chair and so he counterattacked by hiring the country's best lawyer to challenge the legality of the electric chair. In vain, and the first victim of the electric chair, William Francis Kemmler, was executed in 1890.

After a few years, American financier and investment banker John Pierpont Morgan Sr. (1837–1913) got tired of it all. Being powerful and owning a share of the Edison Electric Light Company, he got Edison fired and forced the company to switch to AC. The war of the currents had ended.

The 1,850 MW coal-fired power station in Počerady, Czech Republic. There is speculation that it may be decommissioned in 2025.

stance &
m's Law

One of the fundamental principles in electrical circuits is the relationship between voltage, current, and resistance. This principle had eluded scientists for many years until German physicist and mathematician Georg Simon Ohm (1789–1854) found the solution.

Georg Simon Ohm was one of seven children of which only three made it into adulthood: Georg Simon himself, his brother Martin, and their sister Elisabeth Barbara. The two boys were taught mathematics, physics, chemistry, and philosophy by their father, a locksmith and autodidact who would have loved to be a scientist.

Georg Simon Ohm attended university, but as he preferred playing billiards to studying, his father sent him out into the world to earn a living on his own. And so he did, albeit modestly, moving from place to place as a teacher and private tutor. Eventually, he returned to the University of Erlangen (Germany) and obtained his PhD in 1811, the same year as his younger brother, Martin.

Martin Ohm pursued a career in mathematics, which led him smoothly to a professorship at the University of Berlin. Georg Simon Ohm, however, was not so fortunate. To support himself, he had to continue teaching at low-prestige schools, notably in Bamberg and Cologne. All the while, he dreamt of securing a prestigious position at a university—like his successful younger brother—but he never managed to obtain one.

At this stage in his lonely, unmarried, and unhappy life, Ohm concluded that if he wanted to become a scientist, he had to behave like one. He began conducting experiments and publishing his results. Like André-Marie Ampère and others, he was inspired by the Danish physicist Hans Christian Ørsted, who had discovered in 1820 that a magnetic needle is deflected by a nearby electric current. Ohm, in turn, focused his attention on electricity and magnetism.

Ohm's first paper, published in 1825, investigates the relationship between the decrease in the electromagnetic force exerted by a current-carrying wire and the length of the wire. Other scientists, such as Antoine-César Becquerel (1788–1878, not to be confused with his grandson Henri) and Peter Barlow (1776–1862), were active in the same field, attempting to formulate

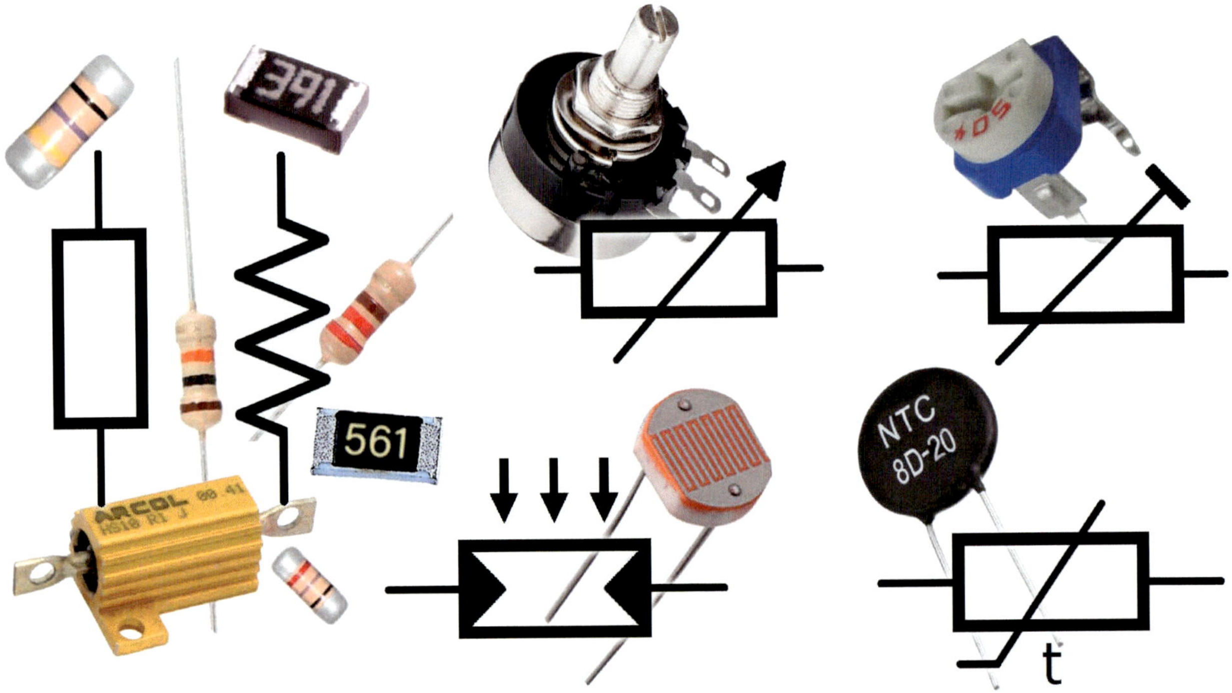

Resistors come in many shapes and sizes, from huge to super tiny. Here are a few examples together with the electric symbols for them. The wiggly-line symbol is the US ANSI symbol, the rectangle-based symbol is the common European (IEC) typ.

The disc-shaped device in the lower right corner is sensitive to temperature; the one on its left is sensitive to light.

a law that accurately described the phenomenon. Barlow came close[4], but Georg Simon Ohm succeeded by introducing the concept of a battery's internal resistance. This brilliant insight not only led to his famous law but also enabled him to solve many unanswered questions that had eluded researchers for years.

In 1827, Ohm presented his theory of electricity in his masterpiece *Die galvanische Kette, mathematisch bearbeitet (The Galvanic Circuit Investigated Mathematically)*. It was quickly embraced by the scientific community—except in Germany, where physics was largely non-mathematical at the time. Despite his growing reputation, Ohm still struggled to secure a university position. His brother's reputation as a 'dangerous revolutionary', following a dispute with the German Ministry of Education, may also have contributed to his difficulties. However, by the end of his life, even the Germans could no longer ignore the significance of his work, and in October 1852, less than two years before his death, he was finally awarded the chair of physics at the University of Munich.

4 Barlow's law is incorrect. However, he is credited with hindering the development of long-distance telegraphy, as he believed he had proved it to be unfeasible.

So, what exactly is Ohm's law? Well, it states that the current flowing through a conductor is directly proportional to the voltage across it and inversely proportional to the resistance of the conductor. This law is mathematically expressed as

$$V = I \cdot R$$

Where R is the resistance in ohms (Ω), V is the voltage in volts (V), and I is the current in amperes (A). (Note that the equation is also often written with a U in place of the V.) It says that for a constant resistance R, if the current increases, the voltage increases too. This equation can also be written as

$$I = \frac{V}{R}$$

This shows that if the resistance increases, the current decreases, assuming the voltage stays the same. In other words, the current is inversely proportional to the resistance of the conductor. Reordering once more, we obtain

$$R = \frac{V}{I}$$

This tells us that when we know the voltage over a conductor and the current flowing through it, we can calculate its resistance.

Before Ohm's discovery, scientists experimenting with the circuit above obtained

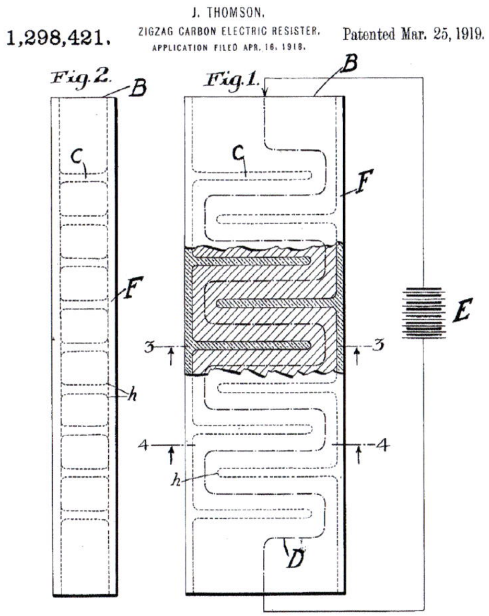

Is this patent for a zigzag carbon resistor (resister) the origin of the ANSI electric symbol for the resistor? No, it isn't, but it could have been.

results that were difficult to capture in a simple first-order equation. In itself, that's not really a problem, it just makes things less simple as hoped for. However, because the results depended highly on the quality of the battery, it was also complicated to develop a more complex theory that fitted all situations.

If the current through the resistor is inversely proportional to the resistor's value, then the results should look like the graph below.

Georg Simon Ohm solved the annoying problem in an elegant way by adding a second resistor in between the battery and the first resistor, as shown in the circuit below.

The resistor colour code is a system used to indicate the value of a resistor using coloured bands printed on the component's body. Each colour corresponds to a number from 0 to 9. The most common format uses four colour bands: the first two represent significant digits, the third is a multiplier (a power of ten), and the fourth indicates tolerance (the allowable variation in value). For example, the code yellow, violet, red, and gold corresponds to 4 (yellow), 7 (violet), ×100 (red), with a ±5% tolerance (gold), giving a value of 4,700 ohms or 4.7 kΩ with a 5% tolerance. Additional bands can be added to increase precision. For instance, five-band resistors are commonly used for 1% tolerance components.

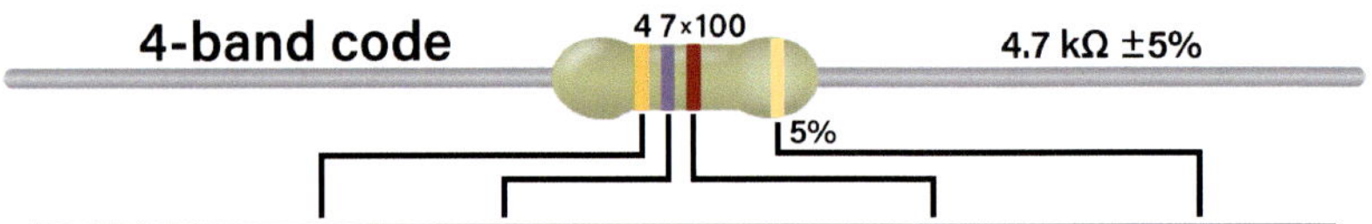

Colour	Band 1	Band 2	Band 3	Multiplier	Tolerance
None	–	–	–	–	±20
Pink	–	–	–	0.001	–
Silver	–	–	–	0.01	±10
Gold	–	–	–	0.1	±5
Black	0	0	0	1	–
Brown	1	1	1	10	±1
Red	2	2	2	100	±2
Orange	3	3	3	1,000	±0.05
Yellow	4	4	4	10,000	±0.02
Green	5	5	5	100,000	±0.5
Blue	6	6	6	1,000,000	±0.25
Violet	7	7	7	–	±0.1
Gray	8	8	8	–	±0.01
White	9	9	9	–	–

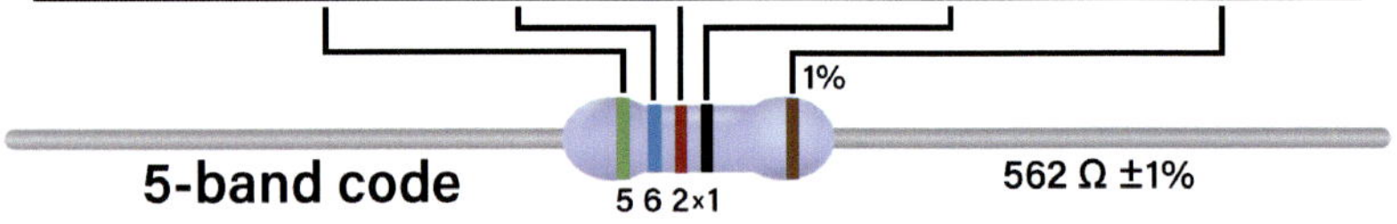

The resistor colour code was originally standardised in the 1920s by the Radio Manufacturers Association (RMA) in the United States, as manufacturers needed a reliable way to mark resistors without print-

Missed Opportunity

English-French natural philosopher and scientist Henry Cavendish (1731–1810) formulated Ohm's law as early as 1781, almost 50 years before Georg Ohm, but he didn't publish his results. Had he done so, not only the unit for resistance would have become the 'cav' instead of the ohm, but smartphones would have been around since 1960, and we would be walking on Mars right now.

ing numbers, which were difficult to read on small parts. In 1952, it was adopted as an international standard by the International Electrotechnical Commission (IEC) and, since 1963, has also been published as EIA RS-279. The colour coding system was later extended to capacitors as well. Today, the international standard for marking resistors and capacitors is defined in IEC 60062:2016. Despite the widespread use of surface-mount components and digital multimeters, the resistor colour code remains a fundamental and widely taught skill in electronics.

The Zero-ohm Resistor

Resistors are used in circuits to control current flow. Yet a commonly used resistance value is 0 Ω, a resistance that does not limit current at all. However, surprising as it may be, it is a very practical device indeed. The zero-ohm resistor looks like any other resistor and has several purposes.

Zero-ohm resistors can be used as jumpers or wire bridges, allowing a trace to cross over other traces. This helps to simplify printed circuit board (PCB) design while making manufacturing easier. Wire jumpers, which might require different equipment during board assembly, can be processed together with the other resistors when using zero-ohm resistors. Finally, zero-ohm resistors are also used to create selectable board options. By adding or removing a zero-ohm resistor, a feature can be enabledç or disabled. This is also useful for testing and debugging, as it allows for isolating subcircuits.

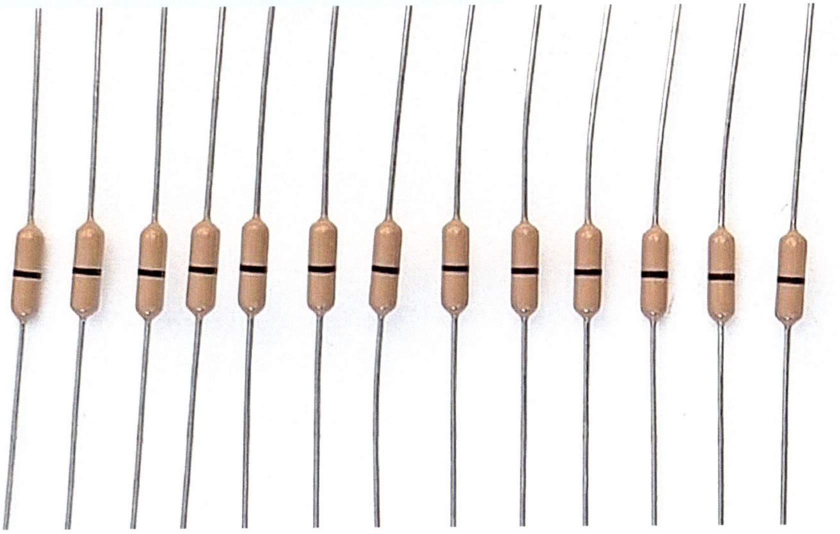

The zero-ohm resistor is an efficient, low-cost solution for circuit routing, configuration, and production flexibility. Shown are large ones that measure about 10 mm, they can be much, much smaller.

Inductance & Electromagnetism

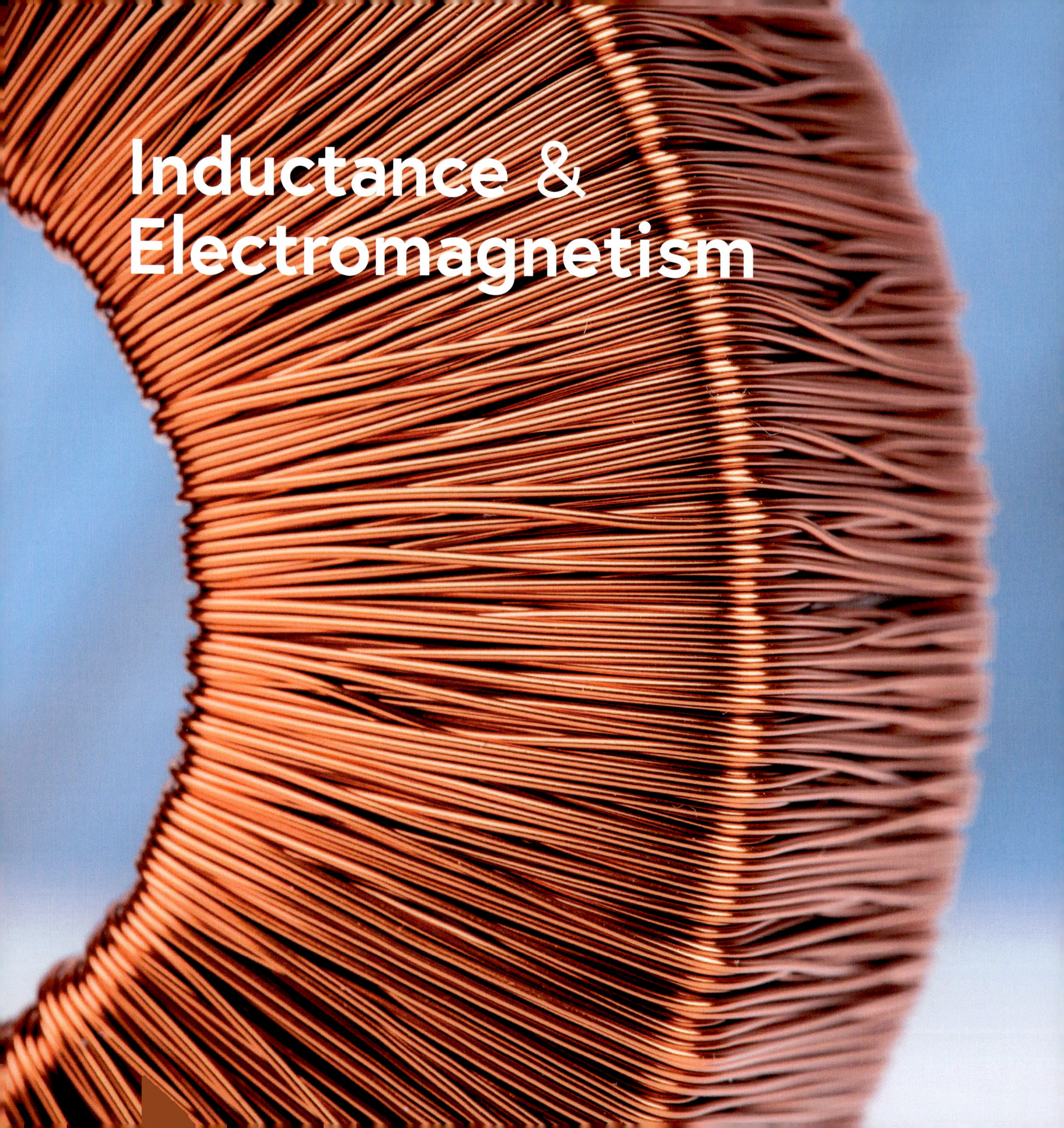

Here is a fun experiment that you can easily do at home. Take a 1.5 V AA(A) battery and one or two small magnets like those neodymium magnets used in all sorts of gadgets to stick something to a fridge. Furthermore, you'll need a thin iron screw or nail with a length of 3 cm so. The length has no real importance, but it must be lightweight and be attracted by a magnet. You'll also need 10 cm or so of thin copper wire. Make sure it is copper as it should not be magnetic. Finally, you'll need some adhesive tape and a bit of thin paper.

Stick the magnet(s) on the head of the screw or nail. Stick the piece of paper on the magnet (not in between the screw and the magnet). Check the weight of this assembly by hanging the pointy side of the nail from the bottom side of the battery. If the nail falls off, it is too heavy. If it doesn't want to remain vertical, then the nail is too short.

If needed, strip a bit of the insulation from both ends of the copper wire. With adhesive tape, stick one end of the wire to one side of the battery. It doesn't matter which end of the battery.

Hold up the battery vertically with the wire on top and hang the nail with its pointy side from the bottom side of the battery, as before. Now, using your free hand, carefully touch the magnet with the free end of the copper wire. If all is well, the nail or screw will start spinning pretty fast. The piece of paper helps to see how fast it is spinning.

Congratulations, you have made an electric motor!

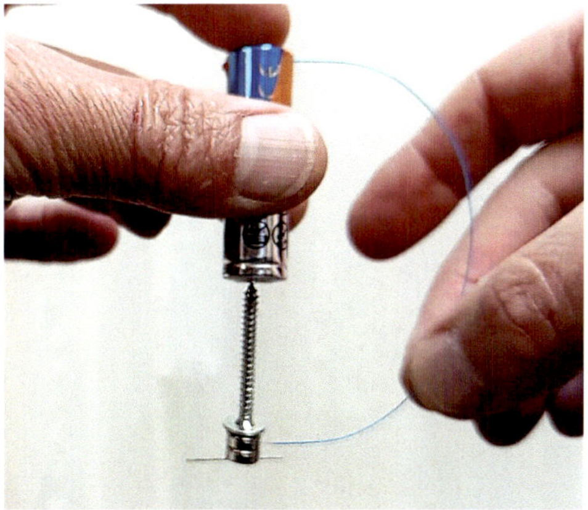

My DIY homopolar electric motor. I used a piece of metal wire instead of paper to show that it was spinning. Be careful though, as the battery will get hot when operated continuously.

Although constructed differently, the motor described above is what Michael Faraday demonstrated at the Royal Institution in London in 1821. As far as we know, it was the world's first electric motor. Faraday called it 'electromagnetic rotation'.

English physicist and chemist Michael Faraday (1791–1867) had started experimenting with magnets and electricity after Danish physicist and chemist Hans Christian Ørsted (1777–1851) had published his discovery of electromagnetism in 1820. Ørsted had discovered that a compass needle is deflected from the magnetic north by an adjacent electric current. Further investigations showed him that an electric current flowing through a wire produces a circular magnetic field around the wire.

Not only Faraday, but many other scientists were triggered by Ørsted's discovery and started to experiment with electromagnetism. One of them was French physicist and mathematician André-Marie Ampère (1775–1836) who set out to develop a mathematical and physical theory to understand the relationship between electricity and magnetism. During this work, he discovered that two parallel wires carrying electric currents will either attract or repel each other. This behaviour depends on the direction of the currents—if they flow in the same direction, the wires attract; if they flow in opposite directions, they repel each other. Ampère's work became the foundation of electrodynamics and, in honour of this, the SI unit of current was named the ampere (A).

Around 1860, Scottish physicist and mathematician James Clerk Maxwell (1831–1879) became interested in electromagnetism. Neither Ørsted nor Faraday had developed a mathematical theory to explain their observations. Ampère, on the other hand, had done some work. Maxwell, being a very skilled mathematician, quickly published some results. A few years later, in 1865, he published his 'Dynamical Theory of the Electromagnetic Field', a work that unifies electricity, magnetism and light. In it, Maxwell demonstrated that all three phenomena travel through space as waves at the speed of light. His theory led to the prediction of e.g. radio waves and X-rays and is one of the main pillars of modern-day electronics. Today, the electromagnetic force is considered one of the four fundamental forces of Nature[6].

Maxwell's theory can be summarized in four equations commonly referred to as Maxwell's equations. They describe how electric and magnetic fields propagate and interact. Originally, Maxwell had formulated twenty equations but English mathematician and physicist Oliver Heaviside (1850–1925) reformulated twelve of them in the form commonly used today.

In these equations, E is the electric field, B is the magnetic field, J is the current density, ρ is the electric charge density, ε_0 is the

6 The other three are the gravitational force, and the strong and weak nuclear forces.

vacuum permittivity and μo is the vacuum permeability.

$$\nabla \cdot E = \frac{\rho}{\varepsilon_0}$$

$$\nabla \cdot B = 0$$

$$\nabla \times E = -\frac{\partial B}{\partial t}$$

$$\nabla \times \mathrm{B} = \mu_0 J + \mu_0 \varepsilon_0 \frac{\partial E}{\partial t}$$

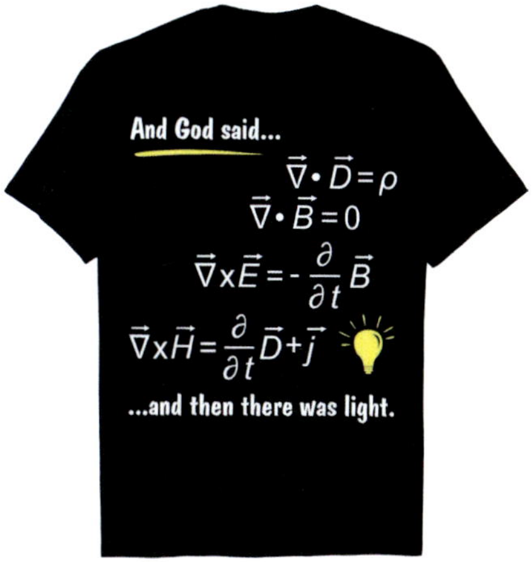

Maxwell's equations are so important that they are being printed on t-shirts.

The first equation is Gauss's Law for Electric Fields. In simple terms, it states that the strength of an electric field depends on the amount of charge in a region.

The second equation is Gauss's Law for Magnetism, and it states that magnetic field lines always form closed loops, implying that magnetic monopoles do not exist.

The third equation, Faraday's Law of Induction, describes how a changing magnetic field over time induces an electric field. This is the principle behind electrical generators and transformers.

Lastly, the fourth equation is Ampère's Law corrected (or extended) by Maxwell. The second part concerning electric charge was added by Maxwell. The equation implies that both a wire carrying a current and a changing electric field produce a magnetic field, enabling the propagation of electromagnetic waves (like light).

It may seem obvious today (doesn't it?), but Maxwell's extension of Ampère's Law led to the discovery that electric and magnetic fields can propagate together as electromagnetic waves moving at the speed of light. They are the basis of all light, radio waves, and other forms of electromagnetic radiation. Heinrich Hertz was the first to confirm Maxwell's equations by experimentally proving that electromagnetic waves exist and propagate through space.

The impact of Maxwell's unification work is enormous. All forms of radio communication, from TV to satellite, radar and microwave oven to Wi-Fi and mobile phone, are explained by it. So are ultraviolet, visible and infrared light (used by e.g. remote controls), or MRI, X-rays and Gamma-rays in medical and scientific imaging. From LEDs and lasers and even particle accelerators to power generators, transformers and electric motors; from climate modelling, finite element analysis (FEA), and quantum electrodynamics (QED) to plasma physics, it is hard to find a field of engineering where electromagnetism does not play an important role.

From a purely electronic-circuit point of view, the inductor is the most important electromagnetic device. An inductor stores energy in a magnetic field when an electric current flows through it. When an inductor is placed in a constantly changing magnetic field, a current is generated in the inductor. This is called induction. Inductance (L) is measured in henries (H), named after American scientist Joseph Henry (1797–1878) who discovered the electromagnetic phenomenon of self-inductance and, independently of Faraday, mutual inductance. Henry is also credited with the inventions of the electric (door)bell (1831) and the electric relay (1835).

The inductor is one of the three passive components that every electronics textbook mentions in one of its first chapters. In electronics, a passive component is a device that does not require an external power source to operate and does not actively control electrical current. Instead, it responds in a predictable way to the flow of current through it or to the voltage over it. Passive components do not generate power, but they can dissipate, store, or release energy. The other two passive components are the resistor and the capacitor[7].

An inductor is a simple device as it is nothing more than a length of wire wound as a coil. Therefore, they are also often called coils. A core made of a material with a high magnetic permeability like iron or ferrite may be inserted inside the coil to improve some of its properties. The core material determines how well the inductor can handle magnetic flux. When there is no core, the inductor is said to have an air core. Inductors are widely used in electrical circuits for tuning, filtering and energy storage.

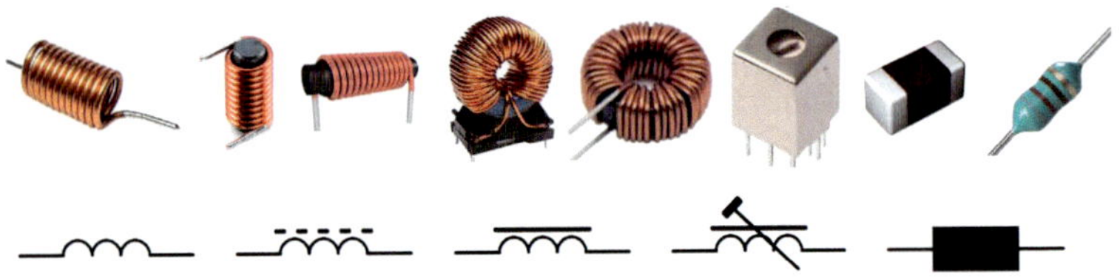

Different kinds of inductors with and without core. The (respective) sizes do not reflect reality. Also shown are several electric symbols for inductors. The symbols with a line parallel to the coil windings are supposed to have a core. The line type refers to the core material.

7 There are more, like the quartz crystal or the antenna, but the resistor, capacitor and inductor are considered the three primary passive components.

Heaviside Had a Way with Words

Besides making Maxwell's equations easier to manipulate, Oliver Heaviside also coined, amongst others, the words 'impedance', 'admittance', 'conductance', 'inductance', 'electret" and 'permeability' that are pronounced every day by electronics engineers all over the world. Furthermore, he invented the step function and introduced the unit impulse function, known today as the Dirac delta function. These functions are heavily (pun intended) used in electronics to characterise the behaviour of circuits.

As said before, a current flowing through an inductor will produce a magnetic field around the inductor. When the current is constant, the magnetic field will be constant too. The magnetic field created this way can be used to attract metal parts or push away other magnets. Inductors used this way are called electromagnets and they are commonly found in e.g. relays and electric door locks. The toaster you use every morning at breakfast uses electromagnetism to keep the bread down. Electromagnets are also used to suspend and drive magnetic levitation trains; its applications are divers indeed.

A loudspeaker too is an electromagnet as it is just a coil with some stuff added to it. Placed in a static magnetic field created with a normal magnet, a variable current like an audio signal flowing through the loudspeaker coil will make the coil move because the magnetic field it creates is attracted by one of the poles of the static magnet (and repelled by the other). If you glue a lightweight (paper) horn to the coil, the horn will move the air around it, and audible sound is produced. But electromagnets also work the other way around. If, instead of pushing a current through the loudspeaker coil, you move the paper horn, a current will be induced in the coil and a voltage develops on its terminals. You have created a microphone.

The loudspeaker is an example of a linear (electric) motor. It uses an electric current to create linear movement. By placing several magnets in series while alternating their orientation (poles), you can move objects. A magnetic levitation train together with its track is basically a very long linear motor. If you place the magnets in a circle around the object, the object will loop around, and the motor becomes a rotary motor. Now an electric current is converted into a rotational movement. This is what we commonly call the electric motor. They

An old bicycle and some electromagnetism let you turn toast into electricity to power the toaster with.

are a bit more elaborate than explained here, but you get the idea.

Like a loudspeaker, an electric motor too can be used the other way around. If you spin its axle instead of driving the motor with a current, it becomes an electric power generator. This lets you convert wood, fuel, water and wind energy into electricity with the help of steam or combustion engines, or mills.

When the current flowing through the inductor is not constant, like an alternating current (AC), the magnetic field will be variable too. Such a magnetic field creates a current in another inductor nearby through induction. The generated current is a copy of the first and produces a voltage between the two terminals of the inductor, which is a copy of the voltage applied to the first inductor. This principle is used in transformers. Depending on how the inductors are made, a transformer creates a smaller, identical or larger copy of the voltage at its input conductor (called the primary, the output is the secondary).

The magnetic field allows transferring energy (power) from the primary inductor to the secondary without an electrical connection. Therefore, a transformer can be used to isolate two circuits. Besides copying a voltage, this is one of its most common applications. The amount of power a transformer can deliver is limited by the magnetic field, while the magnetic properties (permeability) of the medium between the two transformer inductors introduce losses. Therefore, a transformer cannot deliver more power than that is put into it, only less.

The voltage-current relationship of an inductor is given by the following equation:

$$V(t) = L\frac{dI(t)}{dt}$$

Here, V is the voltage over the inductor (in volt), t is the time (in seconds), L is the inductance (in henry), and I is the current through the inductor (in ampere). This equation states that the voltage over an inductor is proportional to the rate of change of the current that flows through it. In other words, an inductor opposes a changing current. It behaves like a resistor for currents that vary in value (e.g. AC). A constant current (which is not a synonym for DC), on the other hand, flows right through it as if it wasn't even there.

Let's conclude this chapter with electromagnetic sensors. Electronic applications often use these to measure the position of an object, the speed of rotation of an axle or the direction of a vehicle. The mechanical compass is probably the most well-known example of a magnetic sensor, but as it doesn't produce any electric current, it is complicated to use in an electronic device[8].

Inductors or coils can be used, of course, but there are more subtle techniques available. One of them is the Hall effect, named after the American physicist Edwin Herbert Hall (1855–1938). Remember that electric current is the flow of charged particles. Also recall that Maxwell's equations relate electric charge to magnetic fields. Now, one of its effects is that when a current flows through a conductor (or semiconductor) and a magnetic field is applied at right angles to the current, the charged particles in the conductor experience a force due to the magnetic field. This is known as the Lorentz force, named after Dutch physicist Hendrik Antoon Lorentz (1853–1928). The Lorentz force causes the charged particles to deflect to one side of the conductor. Charge accumulates on one side of the conductor while a deficit is created on the other. As a result, a measurable voltage develops across the conductor. This voltage, the Hall voltage, is proportional to the strength of the magnetic field and the current flowing through the conductor.

When the charged particles deflected by the Lorentz force move to the side of the conductor, their path through the conductor changes too. The particles take longer or more difficult routes through the material, which increases the resistance. The deflection is proportional to the strength of the magnetic field, and the resistance increases accordingly. This is called magnetoresistance. Iron, cobalt and nickel are materials that exhibit magnetoresistance and so does graphene.

Both magnetoresistors and Hall-effect sensors are commonly used in electronic circuits to detect magnetic fields.

8 A microelectromechanical system or MEMS can do this.

Kirchhoff's Circuit Laws

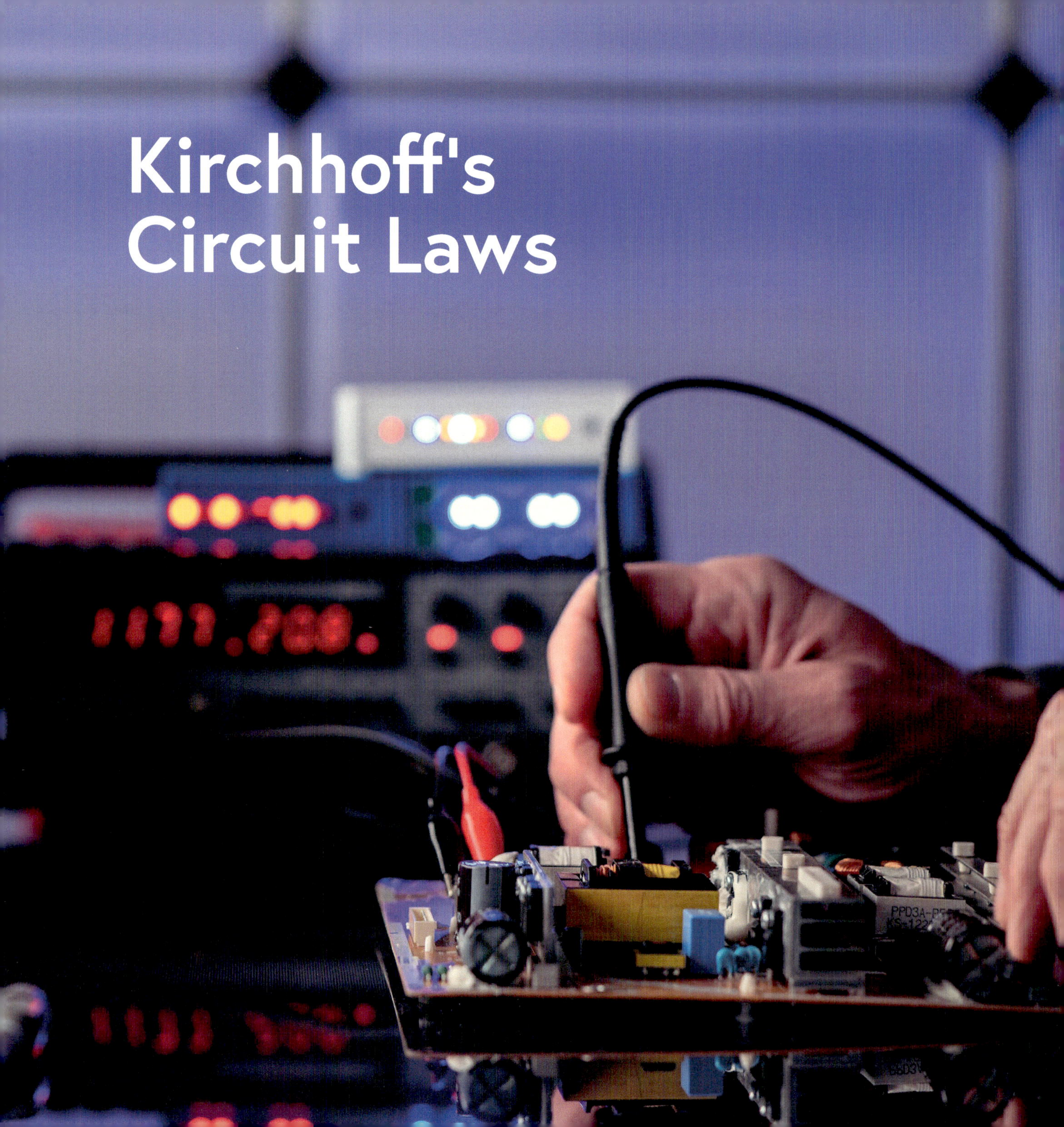

When you ask a scientist about the works of German physicist Gustav Robert Kirchhoff (1824–1887), chances are that he or she will mention spectroscopy and black-body radiation. And (s)he would be right. Kirchhoff coined the term black-body radiation and, together with chemist Robert Bunsen, pioneered spectroscopy to discover new elements like caesium and rubidium. However, an electronics engineer answering the same question will probably only mention Kirchhoff's circuit laws. Kirchhoff formulated his circuit laws in 1845 while he was still a student. He then spent the rest of his life on other subjects like thermal radiation and spectroscopy. Even though his circuit laws were only the subject of his dissertation, they did become the basis of modern electrical circuit analysis. His laws or rules generalize the work of Ohm and preceded the work of Maxwell.

There are two circuit laws: Kirchhoff's Current Law and Kirchhoff's Voltage Law. They are easy to understand and are essential for calculating currents and voltages in a circuit.

Kirchhoff's first circuit law (or rule) is the Current Law. It states that the total current entering a junction (or node) in a circuit is equal to the total current leaving the junction. This means that electric charge cannot accumulate at a node. The law is based on the principle of conservation of charge. Mathematically, it can be written as:

$$\sum I_{in} = \sum I_{out}$$

Or at a node:

$$\sum_{i=1}^{n} I_i = 0$$

where I represents current.

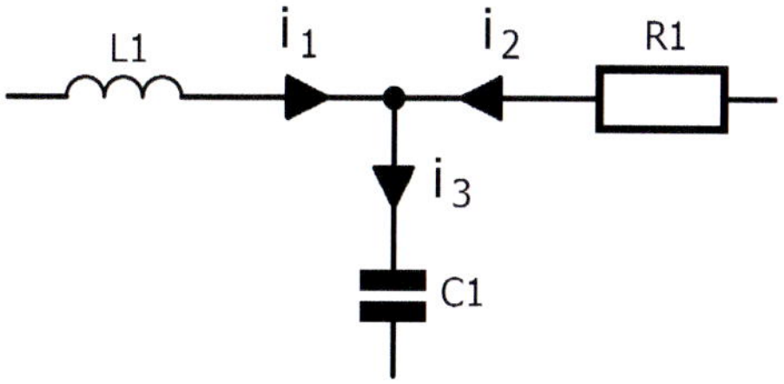

Kirchhoff's current law states that the current entering any junction is equal to the current leaving that junction. Therefore, $i_1 + i_2 = i3$.

Kirchhoff's Voltage Law states that the sum of all voltages around any closed loop in a circuit is zero. In other words, the total energy gained and lost around a loop must be zero. This is based on the principle of energy conservation. Mathematically, it can be written as:

$$\sum_{i=1}^{n} V_i = 0$$

where V represents the voltage drops or rises in the loop.

According to Kirchhoff's voltage law, the sum of the voltages around the loop must be equal to zero. Therefore, $V_1 + V_2 + V_3 + V_4 = 0$.

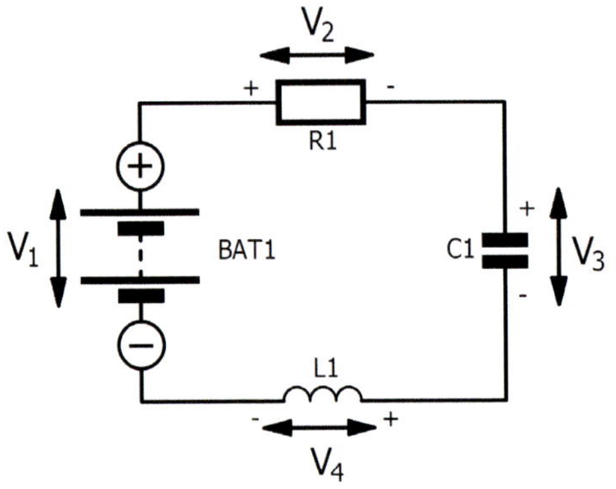

Let's try an example. The figure below shows a simple circuit where the state of LED1 is controlled by pushbutton S1. The circuit has three loops and, when pushbutton S1 is closed, three currents will flow. With the help of Kirchhoff's circuit laws, we can calculate their values.

A simple circuit to switch an LED on and off with a pushbutton (S1). 33k means 33,000 Ω, 330R means 330 Ω.

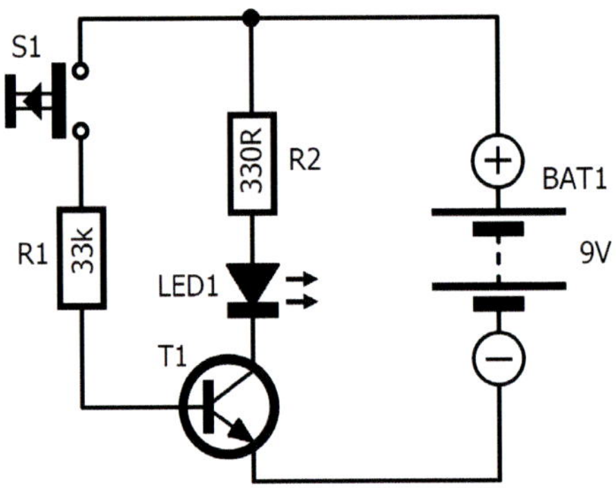

A little bit of extra information concerning the behaviour of LED1 and T1 is required to do the calculations. In this example, when a current flows through LED1, even a tiny current, its so-called forward voltage drop (V_{FWD}) is 1.8 V. This is an acceptable approximation of a red LED. When transistor T1 is in saturation, its base-emitter voltage (VBE) is assumed to be 0.7 V, and its collector-emitter voltage (V_{CE}) is 0.1 V. Again, an acceptable approximation of a standard silicon NPN transistor. The figure below shows all the information.

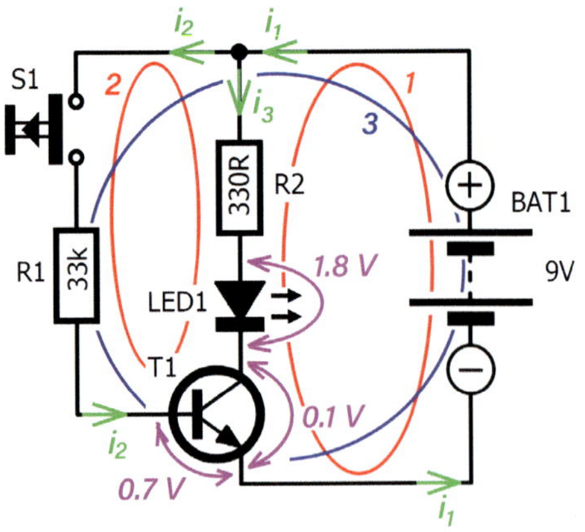

The example circuit completed with the missing information for T1 and LED1. There are three loops. The currents i1, i2 and i3 can now be calculated and also the voltages over R1 and R2.

Suppose S1 is being pressed down. Following Kirchhoff's Voltage Law for loop 1, we can write for the voltage across R2:

$$9 - V_{R2} - 1.8 - 0.1 = 0$$

$$V_{R2} = 7.1\ V$$

Similarly, the voltage across R1 is given by (loop 3):

$$9 - V_{R2} - 1.8 - 0.1 = 0$$

$$V_{R2} = 7.1\ V$$

Using Ohm's law, we can now calculate currents i_2 and i_3:

$$i_2 = \frac{V_{R1}}{R1} = \frac{8.3}{33000} = 2.51 \cdot 10^{-4} A = 251\ \mu A$$

$$i_3 = \frac{V_{R2}}{R2} = \frac{7.1}{330} = 0.0215\ A = 21.5\ mA$$

From Kirchhoff's Current Law we obtain i_1:

$$i_1 = i_2 + i_3 = 0.0218\ A = 21.8\ mA$$

When S1 is released, transistor T1 stops conducting. This is the same as removing T1 from the circuit. Now, all the loops are broken, and no current will flow.

In the previous analysis, we did not use loop 2 because we don't know the voltage between the base and the collector (VBC) of transistor T1. But now that we have done the calculations, we can deduce this voltage from the results obtained above:

$$V_{R1} + V_{BC} - V_{R2} - V_{LED1} = 0$$

$$8.3 + V_{BC} - 7.1 - 1.8 = 0$$

$$V_{BC} = 0.6\ V$$

In practice, Kirchhoff's circuit laws are usually used the other way around. With that I mean that instead of determining the voltages and currents in a circuit from the values of the components making up the circuit, the values of the components are calculated as a function of the voltages and currents that the designer desires in his or her circuit. To clarify this, let's look at the current through LED1 that we calculated to be 21.5 mA. Suppose that we want to make the battery last longer by reducing the current through the LED. Let's say 5 mA still provides a usable light intensity. So how do we go about?

First note that the voltage over LED1 will be 1.8 V when a current flows through it, as that is how we defined the LED (this is an acceptable approximation as a real LED behaves almost the same). Therefore, Kirchhoff's Voltage Law for loop 1 does not change, and so V_{R2} = 7.1 V. With Ohm's law, we calculate a new value for R2:

Gold From the Sun

Like many 19th-century scientists, Gustav Kirchhoff too was active in several disciplines. Together with German chemist Robert Bunsen (known from the Bunsen burner) he invented the spectroscope, which they used to identify the elements in the sun. This gave rise to the following anecdote as reported in 'A Memoir of Gustav Robert Kirchhoff' written in 1888 by a student of Kirchhoff named Robert von Helmholtz (indeed, son of Hermann[10]).

The question whether Fraunhofer's lines reveal the presence of gold in the sun was being investigated. Kirchhoff's banker remarked on this occasion: "What do I care for gold in the sun if one cannot fetch it down here!" Shortly afterwards Kirchhoff received from England a medal for his discovery, and its value in gold. While handing it over to his banker, he observed: "Look here, I have succeeded at last in fetching some gold from the sun."

$$i_3 = 5\ mA = 0.005 = \frac{V_{R2}}{R2}$$

$$R2 = \frac{V_{R2}}{0.005} = \frac{7.1}{0.005} = 1420\ \Omega$$

Loop 3 did not change, so current i_2 is the same as before. The total current i1 is again calculated with Kirchhoff's Current Law:

$$i_1 = i_2 + i_3 = 5.25\ mA$$

The battery will last four times longer.

In a real-life application, instead of 1420 Ω for R2, we would use a standard value of 1500 Ω. What will the total current consumption be in that case?[9]

9 4.98 mA. Multiply i1 by the ratio 1420/1500 and add i2.

10 Do we know Hermann von Helmholtz? Yes, we do. In electromagnetism, German physicist and physician Hermann Ludwig Ferdinand von Helmholtz (1821–1894) is known for the Helmholtz coil and the Helmholtz equation. His son Robert appears to have died young in 1889 at the age of 27. Who knows what he might have discovered had he been allowed to come to age.

This may be a good moment to briefly introduce electric power, the rate at which electrical energy is transferred or converted into other forms of energy, such as light, heat, or mechanical energy. Electric power P is defined as:

$$P = V \cdot I$$

Here, P is the power in watts, V is the voltage in volts, and I is the current in amperes.

In AC systems, electric power is a bit more complicated because part of it is stored in magnetic and electric fields. This is called the reactive power. Only the real power is used for work. The sum of the reactive and real power is called the apparent power S. To correct for this, the equation above must be rewritten as:

$$S = V \cdot I$$

$$P = V \cdot I \cdot \cos\varphi$$

Here, S is in VA (volt-ampere), and cos φ is the power factor, i.e., the efficiency of power usage. Therefore, the power in watts, the real power, can never be larger than the power in VA. In a DC system, the power factor is equal to one.

The unit of power is named after Scottish inventor and mechanical engineer James Watt (1736–1819) who did not contribute directly to the field of electronics. However, by introducing horsepower to compare steam engine power to horses, he laid the foundation for measuring power scientifically and this is how the watt became the unit of power.

Extreme Kirchhoff

This cube is made from twelve 10 kΩ, 5%, 0.25 W resistors from the same batch. When a 10 V power supply is connected to points A and G, the voltage between the points D and H will be 2 V. What is the voltage difference between points C and E? Answer: 2 V. Simulating this simple circuit may not produce correct results.

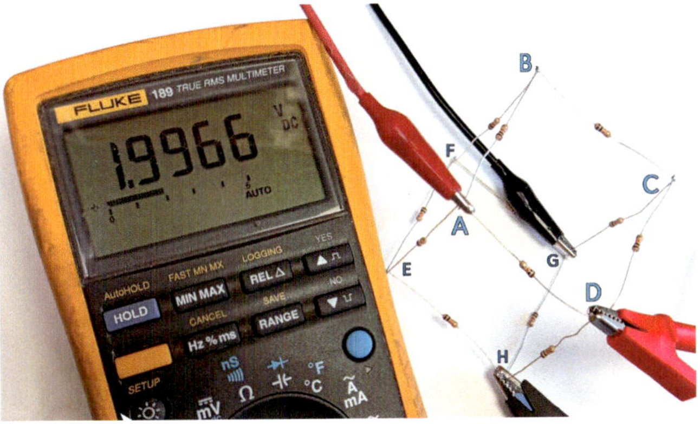

The Vacuum Tube

It is a well-known fact that the first successful transatlantic radio communication has been attributed to by Italian inventor and electrical engineer Guglielmo Marconi (1874–1937). In 1901, by sending the letter 'S' in Morse code (three short clicks or 'dots') over a distance of more than 2,000 miles across the Atlantic Ocean, Marconi proved that radio waves could travel beyond the horizon. Up to then, it was believed that radio signals could only follow a line-of-sight path. A lesser known but nevertheless interesting fact is that it wasn't Marconi who designed and built the radio transmitter used for this feat, but a subcontractor that went by the name of Ambrose Fleming.

English electrical engineer and physicist Sir John Ambrose Fleming (1849–1945), a former student of James Maxwell, had become proficient in power electronics and was hired by the Marconi company to design radio transmitters for transatlantic communication. To improve the detection of the radio signal at the receiver's end, a sensitive device was required. Crystals had been used as radio detectors since 1894, but they were not efficient at the radio frequencies used by Fleming.

In 1880, Thomas Edison had, while trying to improve the incandescent lamp, discovered what is now known as the Edison Effect. Edison and his team had added an electrode inside the lightbulb to divert the charged carbon particles that tended to blacken the lamps on the inside. What they observed was thermionic emission, the liberation of charged particles from a hot electrode. However, as the electron hadn't been discovered yet, they didn't know how to make best use of the phenomenon.

Years later, Fleming, knowing about the electron (discovered in 1897), remembered the Edison Effect and applied it to his problem of detecting radio waves. He constructed a vacuum device capable of conducting current in only one direction, exactly like a fluid valve. Fleming had invented the thermionic diode[11]. He named his invention the oscillation valve, but that didn't stick. Instead, it became known as the Fleming or thermionic valve and as vacuum tube.

Missed Opportunity

For the transatlantic radio experiment, Marconi chose to send the letter 'S' in Morse (dit-dit-dit or '• • •'). However, Ambrose Fleming preferred the letter 'V' (dit-dit-dit-dah or '• • • –'). The official reason is unknown, but I am convinced it was for added dramatic effect. Indeed, if you whistle or sing "dit-dit-dit-dah", you will recognise the iconic four-note motif that opens the first movement of Beethoven's Fifth Symphony: ta-ta-ta-TAAAA

In a thermionic diode, the signal is applied to an electrode, the cathode, inside a vacuum tube. The signal heats up the cathode, which, as a result, starts to liberate electrons. This is called thermionic emission. A second electrode, the anode, is placed at some distance from the cathode. When the signal makes the cathode negative with re-

11 The term 'diode' was coined in 1919 by William Eccles.

spect to the anode, the freed electrons will move to the anode and a current starts to flow. However, when the signal makes the cathode positive with respect to the anode, the electrons won't move to the anode, because it repels them. No current can flow. Therefore, current can only flow if the input signal has the right polarity. In electronics, this is called rectification.

Note how the electric symbol for the diode symbolizes the unidirectional current flow. In this figure, the anode is on the left while the cathode is on the right.

Fleming's valve was an important step forward in electronics. American inventor and electrical engineer Lee de Forest (1873–1961) started to experiment with it

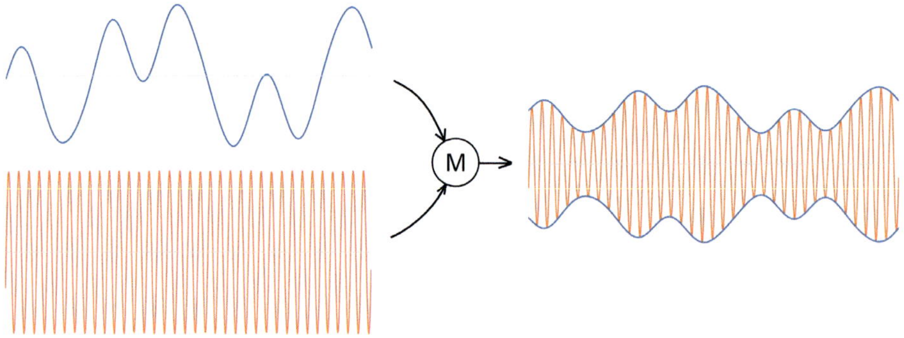

The basic principle of radio communication is quite simple. It starts with the signal to transmit. This signal is superimposed on a high-frequency carrier using a modulation method, which can be as simple as multiplying the carrier with the signal (this is called amplitude modulation or AM). The modulated carrier can be transmitted with a radio.

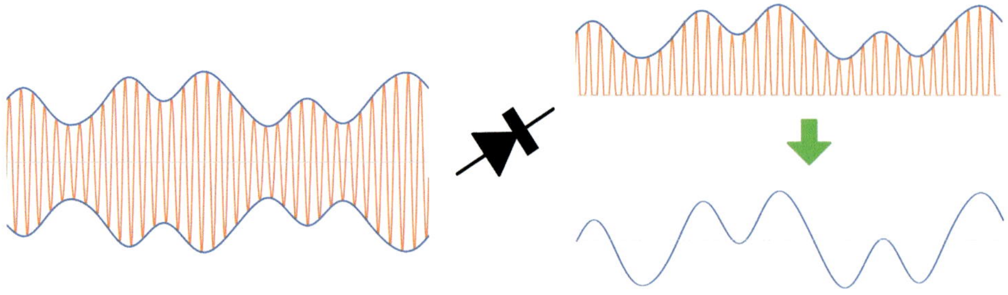

The receiver sees the same signal (but highly attenuated). Simply suppressing the carrier would make the signal completely disappear because the remaining positive and negative sides (the envelopes) add up to zero. However, when the modulated carrier is rectified by passing it through a diode like the one invented by Fleming, the negative part (or positive, depending on the orientation of the diode) of the modulated carrier disappears. Now, when the carrier is suppressed with a filter, only the positive side remains, which is the original signal.

and created various vacuum tube variants that he called Audions. When he inserted a third electrode in between the cathode and the anode, the 'grid' as he called it, he created the first practical electronic amplifier. The invention was patented in 1908 and became known as the audion. The previous audions were forgotten while the last gained fame as the triode.

The audion wasn't a very good amplifier until the American electronics engineer Harold D.[12] Arnold (1883–1933) improved it. The triode and its siblings, tetrode, pentode, hextode, heptode and even octode with fancy names like 12AX7, EL34, EL84 or V69 conquered the world until they were gradually replaced by the transistor starting at around 1960.

The triode works as follows. Like the diode, the triode has a cathode and an anode. However, instead of using the cathode as the device's input as we did before for the diode, it is simply heated to create thermionic emissions. The anode is made positive with respect to the cathode, allowing it to attract the liberated electrons and establish a current. To make this device useful, a third electrode, the grid, is inserted in between the cathode and the anode. As found by Lee de Forest, this third electrode allows to control the intensity of the electron flow.

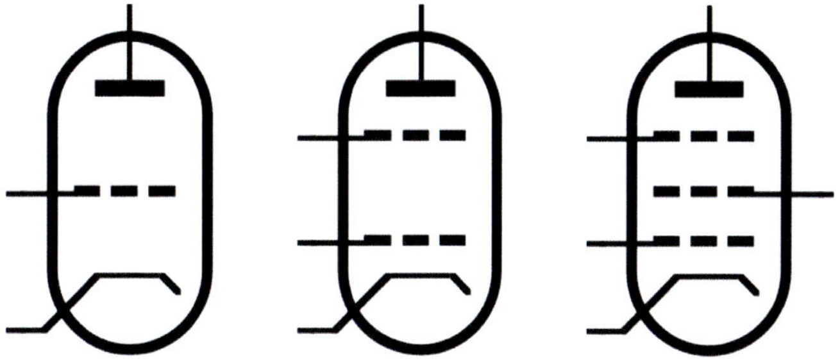

Examples of the electric symbol for a triode, tetrode and pentode (from left to right). The anode is at the top, the cathode at the bottom. Actual symbols may vary depending on the internal structure of the device. The heated electrode is responsible for the thermionic emissions and that typical vintage valve glow.

Therefore, a signal applied to the grid influences the current flowing from the cathode to the anode. When the input signal is small compared to the electron flow, the anode will produce an amplified copy of the input signal. The triode is an amplifier. Adding more grids allows finetuning the control of the electron flow, and this is how we got tetrodes, pentodes, etc.

The current-voltage relationship for a vacuum tube diode is described by the Child-Langmuir law:

$$I_P = k \cdot V_P^{2/3}$$

Here, I_P is the plate (anode) current (in ampere), V_P is the plate voltage relative to the cathode (in volt), and k is a constant that

12 The D in Harold D. Arnold stands for 'DeForest'. What's in a name, eh?

Radio in a Bottle

The Loewe 3NF vacuum tube first made in 1926 integrated an almost complete radio receiver into a single tube. The device contained an amplifier, composed of three triodes, two capacitors and four resistors in a six-pin device. The passive components were each sealed inside a glass tube to avoid them contaminating the vacuum of the tube. It is said (but hard to verify) that this tube was created to avoid a tax levied in the Weimar Republic (Germany from 1919 to 1933) on the number of tubes a radio receiver had. If this is true, then it probably applied to radio manufacturers as radio listeners already had to pay a rather high annual fee (60 mark in 1924). Radios based on the Loewe tube were much cheaper. Loewe also created the 2NF (two tetrodes plus passive components) and the WG38 (two pentodes, a triode and the passive components).

depends on the geometry of the tube and the cathode's electron emission properties.

$$k \cong 2.33^{-6} \cdot \frac{A}{d^2}$$

Here, A is the effective emitting area of the cathode (in square meters) while d is the distance between the cathode and anode (in meters). Because, in this equation, d is squared, it is advantageous to keep the diode small.

Adding a grid to the diode to turn it into a triode doesn't make things easier. There is no easy equation describing the voltage-current characteristics of a triode, but there is an empirical rule that extends the Child-Langmuir law:

$$I_P = k \cdot V_P^{\frac{2}{3}} \cdot \left(1 - \frac{V_G}{V_C}\right)^{\mu}$$

Here, VG is the grid voltage (in volt), VC is the cutoff voltage (in volt), and μ is the amplification factor. For small signals (written in lowercase), the relationship between iP and vG may be approximated by:

$$i_P \approx g_m \cdot v_G$$

$$\mu = g_m \cdot r_P$$

where gm is the transconductance (typically in millisiemens[13]), which is the change in IP per unit change in VG. Its value depends on the tube, and is related to μ:

In this equation, rP is the dynamic plate resistance (in ohm) for a given grid voltage VG. The values of μ, rP, gm and are obtained from the vacuum tube's datasheet or through experimentation.

13 S = 1/Ω. It is also known as 'mho', ohm written backwards.

Much Ado About *Almost* Nothing

The book you are reading right now sort of follows the official Wikipedia opinion and timeline on who invented what and when. Hans Camenzind, creator of the legendary NE555 timer, self-published an apparently well-researched book entitled **Much Ado About Almost Nothing** (2nd edition, 2008) in which he retraces, among others, the real history of the vacuum tube and radio. A recommended read that puts things in the perspective of patent disputes, charlatans, large companies, money, and tragic careers.

See www.historyofelectronics.com

The Transistor

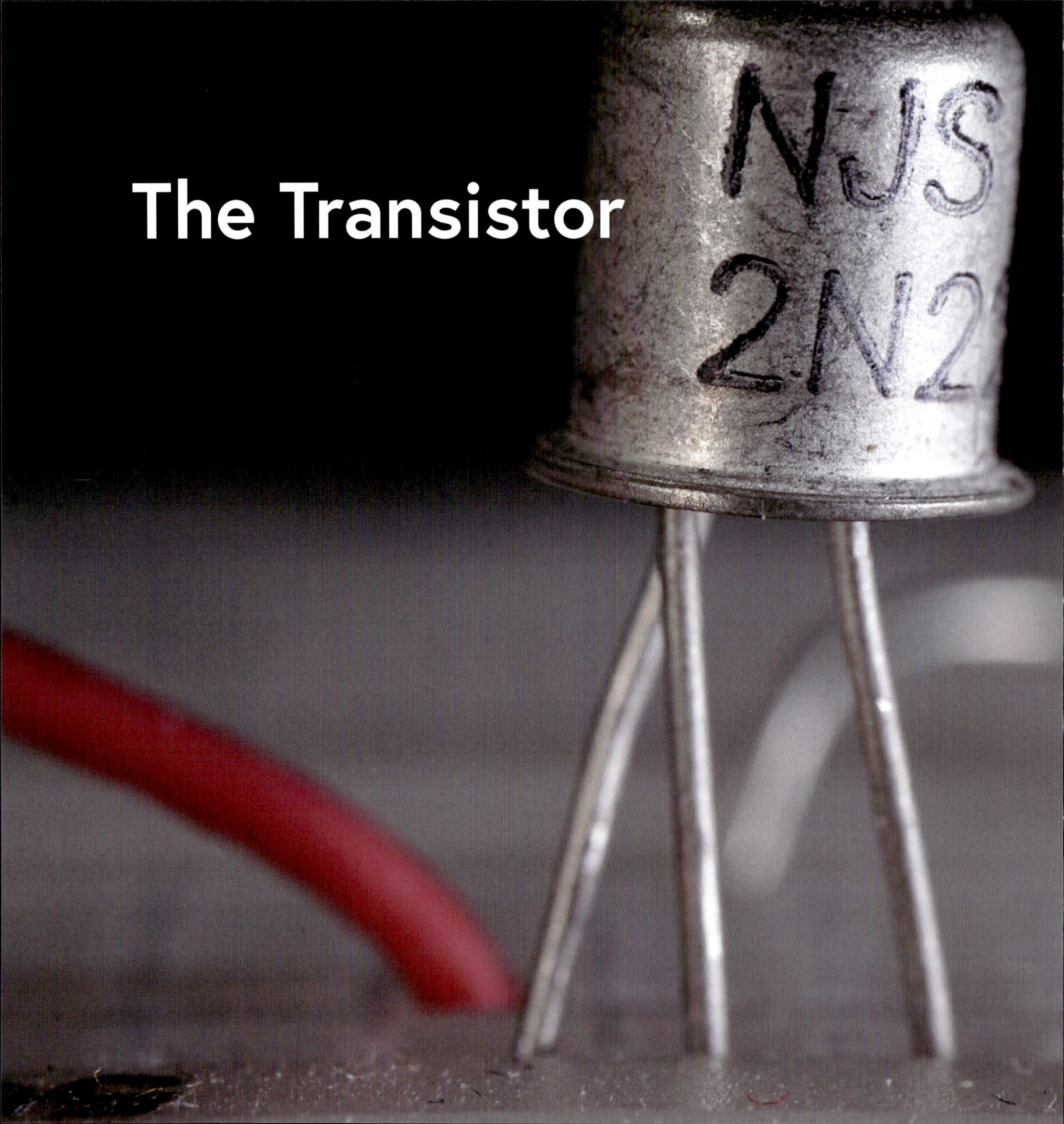

One of the main events that shaped today's electronic world was the invention of the point-contact transistor on 16 December 1947 by American physicists Walter Houser Brattain (1902–1987) and John Bardeen (1908–1991) and, a few weeks later, on January 23, 1948, by the invention of the junction transistor by American physicist William Bradford Shockley Jr. (1910–1989), all working at Bell Labs. This is not a total coincidence as Bardeen, Brattain, and Shockley worked in the same team. Being a relatively recent invention, the process of how this all came about is quite well documented. A lot of information can be found online in transcripts of interviews with and detailed papers written by the people involved in the process that cumulated in the modern transistor. The precision of the dates of the conception of world's first two transistors shows this.

Lesser known, however, is why the modern transistor is made of silicon. "Duh, because silicon has better properties than germanium, okay?" is not the answer we are looking for. What I meant is that, basically, transistors are made of sand. Now where did we get that idea? Sand is an insulator; it is even used in fuses to make them safer. Silicon can be extracted from sand, but it isn't a conductor either and it isn't even a good semiconductor. In its pure form, silicon has a very high electrical resistivity and, under standard conditions, behaves more like an insulator. Using such a material to create an amplifier with seems like a far-fetched idea, to say the least. So, let's try to trace back its history.

Back in the 19th century, electricity was a hot topic, and many scientists conducted all sorts of experiments to try to better understand the phenomenon. Around 1820, German physicist Thomas Johann Seebeck (1770–1831) found that a circuit made from two dissimilar metals with junctions at different temperatures would deflect a compass needle. This effect became known as the Seebeck- or thermoelectric effect. Seebeck also reported that for certain minerals like lead sulfide (PbS, galena) and iron sulfide (FeS2, pyrite) in contact with copper, the effect was much stronger that observed for several of the metals he had tried.[14]

Michael Faraday too studied the electrical behaviour of various materials under different conditions. In 1833, he discovered that the resistance of silver sulfide (Ag_2S, argentite) decreased with increasing

14 Today, Seebeck is considered to be the first to have noticed that semiconductors exhibit special electrical behaviour. Alessandro Volta coined the word 'semiconductor' as early as 1782, but in a different context.

temperature. This was unexpected as the resistance of other (metal) conductors was known to go up with temperature. The common explanation was that heating created more friction to the flow of the 'electric fluid' through the conductor, which increased its resistance. Faraday attributed his strange result to the ionic conduction properties of silver sulfide. He thought that certain materials conducted electricity through the movement of particles rather than by the flow of electric fluid.

Karl Ferdinand Braun, creator of the first semiconductor device, also invented the phased-array antenna, the cathode-ray tube (CRT), and the CRT-based oscilloscope.

In 1874, inspired by an article by German physicist Hermann Herwig (1844–1881), German electrical engineer, physicist and inventor Karl Ferdinand Braun (1850–1918) started experimenting with metal wires connected to metallic sulfide crystals (like galena, pyrite, and chalcopyrite, $CuFeS_2$). He observed unilateral conductivity (i.e. rectification) in a semiconductor and today he is considered the creator of the first semiconductor device. Note that Braun obtained similar results with copper oxide (Cu_2O), a material that became popular in the 1920s for building rectifiers with[15].

In 1895, Indian polymath Sir Jagadish Chandra Bose (1858–1937) demonstrated wireless transmission and remote control. In his experiments, he made use of a galena crystal. This was the first time a semiconductor junction was used as a detector in radio communication. He is the father of the crystal detector (and much more). He was also a firm believer in open source and refused to patent his inventions.

In 1906, American engineer and scientist Greenleaf Whittier Pickard (1877–1956) filed a patent for a 'thermo-electric regenerative detector' that used silicon. He was inspired by an observation made in 1879 by British inventor David Edward Hughes (1830–1900). The latter had noticed that

15 Metal-semiconductor junctions as studied by Braun are now known as Schottky barriers, named after German physicist Walter Hans Schottky (1886–1976). They are used to make Schottky diodes with.

The Cat's Whiskers

In the 1920s, commonly referred to as the Roaring Twenties, "the cat's whiskers" was a popular expression, similar to "the bee's knees," that meant something exceptionally good. For example: "Check out my new radio—it's the cat's whiskers!"

his telephone was affected by distant spark discharges.[16] Pickard interpreted the telephone's carbon microphone as a detector and started experimenting. Not satisfied with the detection quality of the steel-carbon combination, he turned his attention to other materials. Pickard claimed to have tried several thousand different materials and that he found some 250 that could be used to build workable detectors with. He settled on silicon as most effective material, having both high electrical resistance and high thermo-electromotive force. A few years later, in 1911, Pickard patented a device that became known as the cat's-whisker detector. This device, consisting of a thin, springy and pointy metal wire that was pushed into a piece of silicon, became widely used. Pickard was probably the first to use silicon as a semiconductor

The motor driving the research in detectors and rectification was the telegraph. Messages in Morse code could be sent over a wire by transmitting electrical pulses. These are easy to detect with for instance an electromagnet. For wireless telegraphy

16 With this observation and the experiments he conducted afterwards, Hughes anticipated the discovery of radio waves by Hertz and radio telecommunication by Marconi.

however, pulse detection was more complicated. Now, at the time, it was well known that metal filings responded to sparks by clumping together. To 'unclump' them, they had to be shaken or tapped. French physicist Édouard Eugène Désiré Branly (1844–1940) developed a device that used this effect and that became known as the coherer. It did the job but wasn't very fast due to its 'decoherer' mechanism that did the tapping.

The crystal detector, albeit needing mechanical adjustment, did not require decohering and therefore was much faster. It was so fast that it could even be used to decode amplitude-modulated (AM) speech. This opened up whole new avenues to explore and radio applications started to develop quickly.

American scientist Russell Shoemaker Ohl (1898–1987) too worked on detectors and rectifiers. His goal was to make them as fast as possible for use in broadcasting and radar applications. He had tried miniaturizing tubes to speed them up, but, hindered by their limitations, he switched to crystal detectors. At the time, crystals were grown by fusing layers of molten material together. When working with such silicon crystals, he noticed that some of his samples produced a small current when exposed to light. Investigating the effect further, he found that it was due to impurities between melt layers. Impurities had already been suspected to be responsible for rectifier effect in crystal detectors, but in what way was not understood. Ohl found that the impurities made some sections more resistant to electrical flow than others. It were the barriers between areas of different purity that made the crystal work. In 1939, Ohl discovered the p-n junction (where 'p' means positive and 'n' negative). He patented his device in 1941 as a 'light sensitive device', known today as the solar cell.

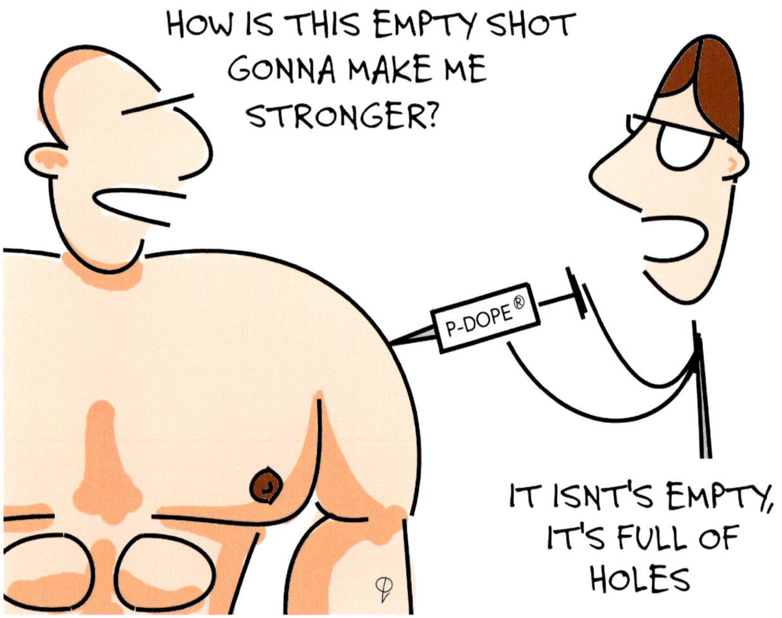

Meanwhile, at Dr. Randy Ohl's office, doping experiments continue...

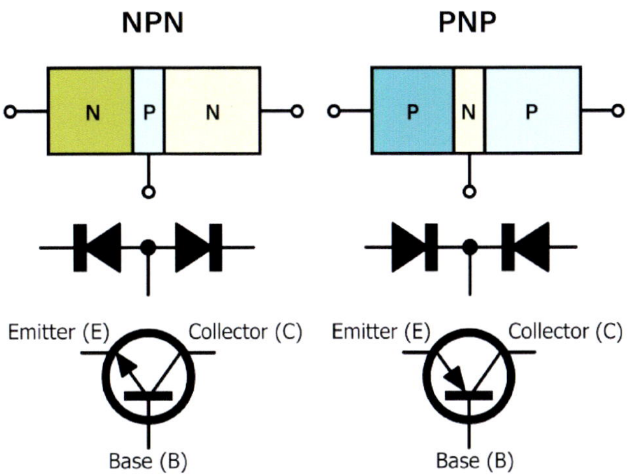

Ohl continued to investigate the p-n junction and the impurities that made it work. He found that by introducing different elements one could obtain either p-type or n-type material. P-type material has free moving positive charges (called holes) whereas n-type material contains free moving negative charges (electrons). This is related to the different number of valence electrons of the material and the impurity (the exact mechanism is beyond the scope of this book). The process of adding impurities to a material is called doping.

On the Origin of 'Transistor'

Initially, Bert Moore, a team member of Bardeen, Brattain, and Shockley, had proposed to call the new semiconductor device the 'persistor' because it required a lot of persistence to produce it. American engineer John Robinson Pierce (1910–2002), the supervisor of the inventors, eventually came up with the name transistor. His reasoning was as follows. As the device was supposed to be the dual of the vacuum tube, which has transconductance, the new device would have 'transresistance'. Furthermore, the name had to fit in with the names of other devices, such as varistor and thermistor. This resulted in 'transistor'.

Walter Brattain (see the beginning of this chapter), a colleague of Russell Ohl, became interested in the p-n layer and wondered if it would be possible to influence the p-n layer by adding an electrode to it. That way, it could be used as an amplifier. This resulted in 1947 in the invention of the germanium point-contact transistor.

William Shockley (see the beginning of this chapter), a team member of Walter Brattain and John Bardeen, didn't like the point-contact much because it was fragile, difficult to manufacture, and its performance was unstable. To overcome these issues, Shockley created a three-layer device where the middle layer could influence the other two. The germanium junction transistor was born in 1948.

The bipolar junction transistor or BJT is created by stacking alternating layers of n-type and p-type semiconductor material. This way, two configurations are possible: NPN (left) and PNP (right). In practice, the layers do not have the same thickness. Each p-n junction can be considered a diode, which is shown by the equivalent diode circuits in the middle. At the bottom are the electric symbols for the NPN and PNP transistors. The circles are often omitted.

A p-n junction is created by stacking a layer of p-doped material on a layer of n-doped material. The assembly works like a diode or a solar cell. If you add a third layer (while alternating the layer types), you obtain either an NPN or a PNP transistor. Adding a fourth layer is, of course, possible and William Shockley did this. This resulted in the Shockley diode. The thyristor or silicon-controlled rectifier (SCR) is a 4-layer device similar to a Shockley diode but with three terminals instead of two (making it a triode). Going beyond four layers is pointless, at least for now. Stacking more than four layers creates too complex and instable devices that are impractical to use.

Remember at the beginning of this chapter we asked ourselves why transistors are made of sand? The first transistors were made of germanium, not silicon. Why? The reasons for this are multiple. First, at the time it was difficult to purify silicon and grow good quality crystals, and therefore silicon was expensive. Because germanium was used a lot, its electrical behaviour was better understood than that of silicon. Furthermore, germanium was easier to dope as it has a lower melting point than silicon. Lastly, germanium devices were faster than their silicon equivalents. However, silicon has better thermal stability than germanium and it is abundantly available in for instance sand. So, when advancements in the silicon purification process allowed silicon to be used for manufacturing transistors, germanium's position weakened. The final blow came with the discovery that silicon can form a stable oxide layer, an insulator, which made it the perfect material to build integrated circuits with (but that is another story). So even though Greenleaf Whittier Pickard had found in 1906 that silicon would be the best material for semiconductors, it took almost fifty years before the first practical silicon transistor saw the light (in 1954).

Now, let's have a brief look at the mathematics of semiconductor devices. Note that what follows is by no means a complete mathematical description. The goal here is simply to give the reader a little insight in the rules that govern p-n junctions. It all starts with the current-voltage relationship for a p-n junction as given by the Shockley diode equation:

$$I = I_S \cdot (e^{\frac{V}{nV_T}} - 1)$$

Here, I is the current through the diode (in ampere), I_S is the saturation current (a very small constant in ampere), V is the voltage across the diode (in volt), V_T is the thermal voltage (a temperature-dependent 'constant', in volt), and n is an 'ideality' twiddle factor. When the diode is being forward biased, i.e. when it is conducting and the power of e is much larger than one, this equation may be simplified to:

$$I = I_S \cdot e^{\frac{V}{nV_T}}$$

This equation shows that the diode current increases exponentially with the voltage applied over it. When the diode is blocking, the power of e goes to zero and the equation may be approximated by:

$$I = -I_S$$

As long as the diode remains reverse-biased, this equation is independent of the voltage over it. This is, however, limited by the reverse breakdown voltage of the diode. When the reverse bias voltage becomes too high, the p-n junction collapses, and the diode starts to conduct again, but in the other direction. So-called Zener and avalanche diodes exploit this effect in a controlled way.

As will be clear by now, the bipolar junction transistor has two p-n junctions and therefore similar equations apply. An approximation for the forward-biased configuration is as follows:

$$I_C = I_S \cdot e^{\frac{V_{BE}}{V_T}}$$

Here, I_C is the collector current and V_{BE} is the voltage over the base-emitter p-n junction. Furthermore:

$$I_B = \beta \cdot I_C$$

$$I_E = I_C + I_B$$

The last equation is Kirchoff's current law; β is the transistor's current gain (also known as h_{FE}), a constant that depends on the transistor and that can range from as low as 20 to as high as 1000. With these equations, we can use the transistor as a current amplifier. In reality, there is much more to consider than presented here, but it is a start into transistor circuit analysis.

The bipolar junction transistor (BJT) marked a new era in electronics. But did you know that the field-effect transistor, better known as the FET, was invented more than 20 years before the junction transistor? It is attributed to the Austro-Hungarian physicist and electrical engineer Julius Edgar Lilienfeld (1882–1963) who filed a patent for it in 1926. Of course, it wasn't called a transistor but a 'method and apparatus for controlling electric currents.' Unfortunately, due to technical limitations, Lilienfeld could not build and try out his invention. Later, William Shockley, unaware of the work of Lilienfeld, reinvented the device in 1939. Shockley had a prototype built out of

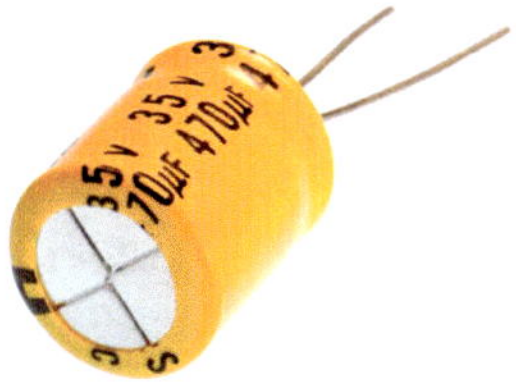

Lilienfield also invented the electrolytic capacitor (in 1931).

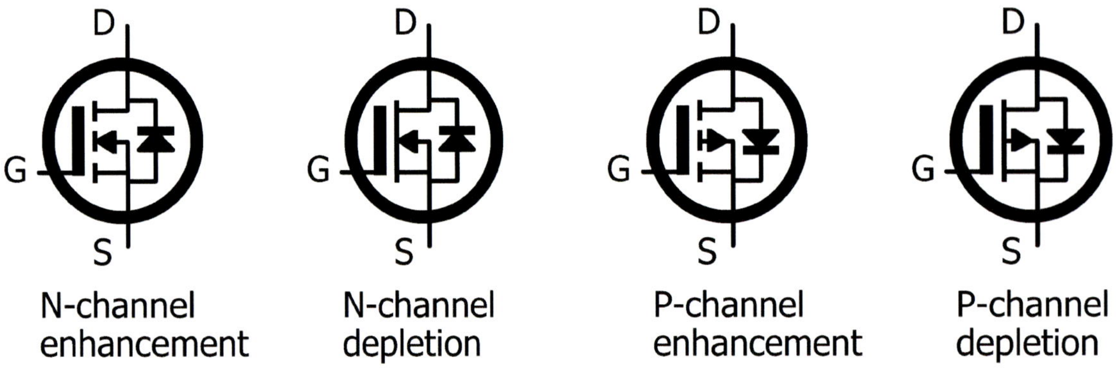

Electric symbols for the MOSFET. 'D' stands for drain, 'G' means gate while 'S' refers to the source.

copper-oxide, but it didn't work. Because of the difficulty of manufacturing FETs, Shockley and his team decided to concentrate their efforts on the bipolar transistor.

Cut-off region	$V_{GS} < V_{th}$	$I_D = 0$
Linear region	$V_{DS} < V_{GS} - V_{th}$	$I_D = 2k \cdot \left[(V_{GS} - V_{th}) \cdot V_{DS} - \frac{V_{DS}^2}{2}\right]$
Saturation region	$V_{DS} \geq V_{GS} - V_{th}$	$I_D = k \cdot (V_{GS} - V_{th})^2$

The first practical FET was created in 1959 with the development of silicon MOSFETs (Metal-Oxide-Silicon FETs) by Korean electrical engineer Dawon Kahng (1931–1992) and Egyptian physicist Mohamed Atalla (1924–2009). The MOSFET is currently the most widely used type of transistor. Maybe someday the invention of the FET will be considered as important as that of the bipolar junction transistor?

The MOSFET can be described mathematically as follows. It features three operating regions:

Here, V_{GS} is the gate-source voltage, V_{DS} is the drain-source voltage, V_{th} is the threshold voltage, I_D is the drain current, and k is a constant that depends on the device. The threshold voltage too depends on the transistor type. In the linear region, the MOSFET behaves like a resistor. Therefore, this region is also called the ohmic region and it can be used for amplifying a signal. The saturation mode is useful in switching applications.

We're Running Out of Sand!

It may be hard to believe, but sand—a primary component in concrete, asphalt, and glass—is actually a finite resource facing a global shortage due to unsustainable extraction rates, driven by rapid urbanisation and infrastructure development. The silicon used in semiconductors is derived from sand. However, not all sand is suitable. Ordinary beach sand, for example, contains too many impurities, making it unsuitable for silicon wafer fabrication. To address the crisis, the electronics industry is exploring recycled silicon from discarded wafers to reduce reliance on raw quartz. Additionally, alternative materials such as silicon carbide (SiC) and gallium nitride (GaN) are being investigated as potential replacements for silicon in semiconductor manufacturing.

Although silicon has been the main semiconductor material since the 1960s, several emerging materials and technologies are being explored for their potential advantages in specific applications. Materials like GaN (gallium nitride), SiC (silicon carbide), graphene, and nanotubes come to mind. These materials feature some unique properties that address limitations of silicon in areas like speed, efficiency, and miniaturization. They will not replace silicon anytime soon but are more likely to complement it in hybrid systems or be adopted for specialized applications such as high-frequency, high-power, or nanoscale devices.

The Integrated Circuit

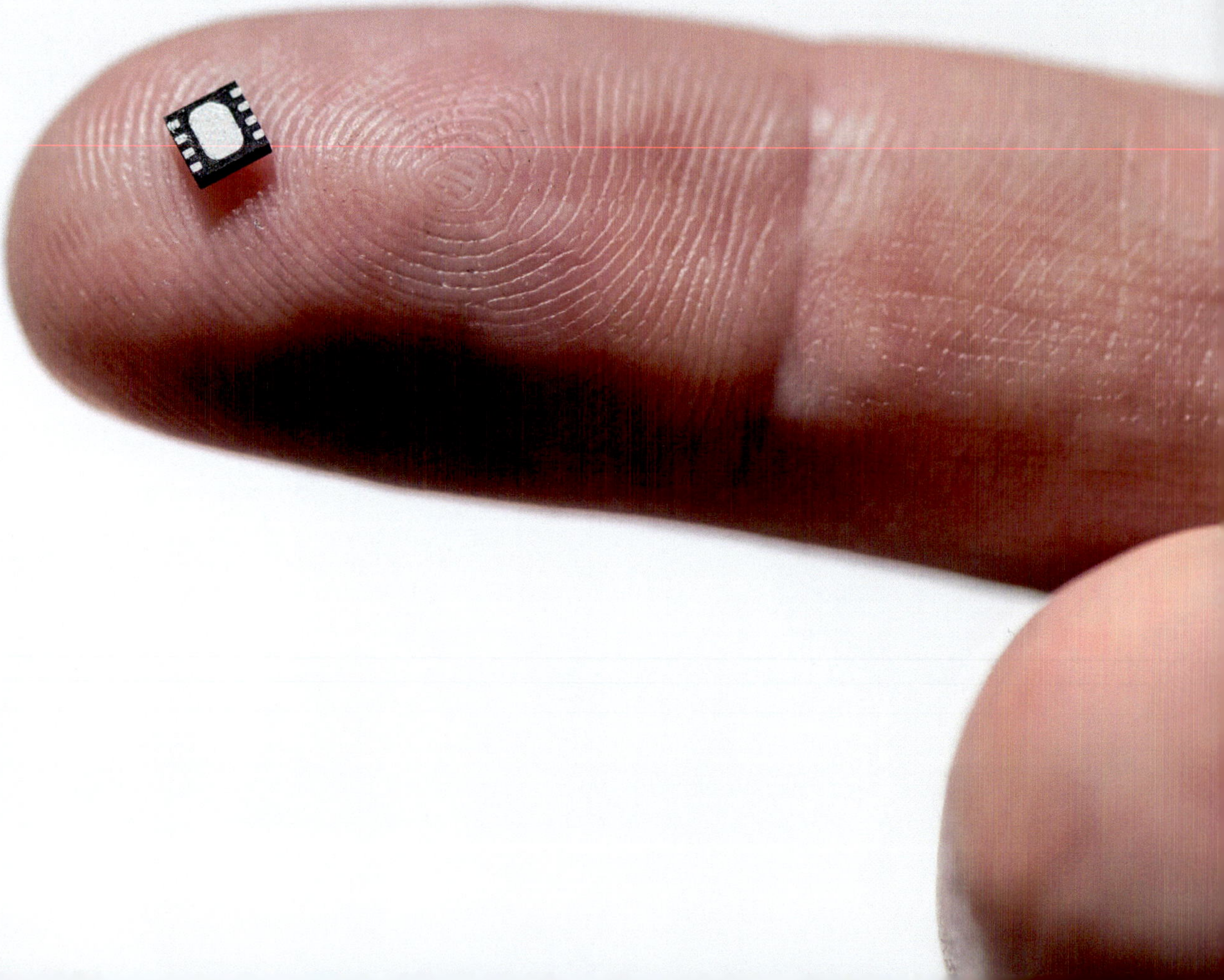

Integrated circuits (ICs), or chips as they are commonly referred to, sure have come a long way since their birth at the end of the 1950s. Today, it is hard to imagine an electronic device without a single IC in it. There exist ICs for almost every imaginable function, from super complex ultra-high-performance microprocessors to simple timers for blinking a light.

An integrated circuit is a circuit built on a small piece of semiconducting material that is able to perform the same functions as the larger circuit that is made of discrete parts. According to JEDEC (Joint Electron Device Engineering Council, a global organization that develops open standards for the semiconductor and microelectronics industry), an integrated circuit or IC is:

"A circuit in which all or some of the circuit elements are inseparably associated and electrically interconnected so that it is considered to be indivisible for the purposes of construction and commerce."

The advantage of ICs over circuits built with discrete components is, besides their much smaller size, their lower cost and better performance. The reason is that the components in an IC are smaller and much closer to each other. This makes them faster and therefore they consume less power. Their costs are low because the components are printed and etched all together instead of being placed on a circuit board one by one. Furthermore, tens or hundreds of identical circuits can be produced on a single wafer in one go. On the other hand, designing an IC is expensive. Building a semiconductor fabrication plant, a foundry, costs even more. As an example, Samsung's 3-nm foundry in South Korea, operational since 2022, has cost roughly $15 billion!

Even though a foundry can produce all sorts of ICs, (very) high production volumes are required to make an IC commercially viable. An IC can be produced more cheaply if its design is compatible with the pro-

Semiconductor-grade silicon used for making integrated circuits has 99.9999999% purity (nine-9 or 9N silicon) and is a nearly defect-free single crystalline material. It is also quite expensive.

duction technology of an older foundry. On the other hand, an IC can become obsolete because the legacy production processes required for it are no longer available.

Why Wafers are Round

Integrated circuits are built as dies on wafers of almost pure silicon. You may have seen a photograph of someone dressed in some kind of space suit holding a shiny wafer. But did you ever wonder why the wafer is round? Is it to keep die-cutting losses to a minimum? Actually, the reason is rather down-to-earth. They are round due to the way they are made. The most common method to produce monocrystalline silicon is the one discovered in 1916 by Polish chemist Jan Czochralski (1885–1953). It consists of dipping a seed crystal mounted on a rod in molten silicon and then pulling the rod slowly upwards while rotating it. When done properly, the result is a single-crystal cylinder. The Czochralski method can be used for semiconductors, metals and salts. Note that rectangular and square wafers exist too

Designing a single IC for a specific application is possible and done often, but only if high volumes can be assured. They are called ASICs for Application-Specific Integrated Circuits. Examples are watches, bike computers, weighing scales and pocket calculators. Inside such products you will usually not find more than one (tiny) ASIC. Such ICs keep the price of a product low, but they are not flexible. Adding or correcting a function requires an expensive redesign. Instead of designing an ASIC for your circuit, you can also build a circuit with general-purpose building blocks. Many semiconductor manufacturers offer a plethora of devices suitable for all sorts of tasks. They are similar to home improvement stores that sell everything from nuts and bolts to garage doors and gardening tools. The operational amplifier (op-amp) is the most common building block IC.

Besides subdividing ICs into ASICs and building blocks, they can also be subdivided into analogue, digital, and mixed signal ICs. Analogue ICs are designed to handle continuous signals, which can vary smoothly over a range of values. They are used to amplify, filter, modify, or otherwise manipulate analogue signals. Examples include op-amps, voltage regulators, filters, and audio amplifiers. Digital ICs, on the other hand, process discrete binary

Inspecting dies on a wafer.

signals, i.e. signals that can take on only two values (zero and one or high and low). A lot of what is inside a computer or smartphone is digital. Microprocessors, microcontrollers, and memory are digital, but also data communication and networking applications use mainly digital ICs. Finally, mixed-signal ICs combine both analogue and digital circuits on a single chip. These ICs are designed to handle both types of signals simultaneously, allowing them to interface with real-world analogue signals and process those signals or convert them into digital data for further processing in digital systems. Conversely, mixed-signal ICs can also convert digital data back into analogue form for output or control applications.

The invention of the integrated circuit (1959) is attributed to the American electrical engineer Jack St. Clair Kilby (1923–2005) and the American physicist Robert Norton Noyce (1927–1990). Added to these two should be the Swiss engineer Jean Amédée Hoerni (1924–1997) who invented the planar process that made IC manufacturing possible. Kilby was the first to integrate a small circuit on a single piece of germanium. The circuit comprised transistors,

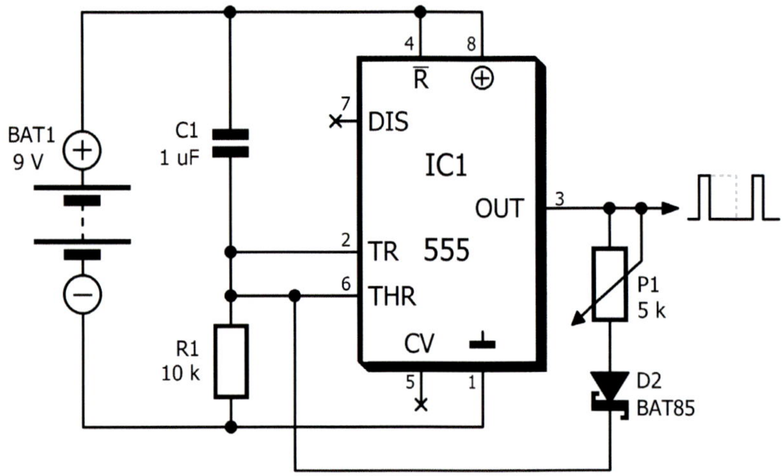

This unusual 555 circuit produces very short positive pulses with a duty cycle down to almost 0%. This is uncommon as most 555 circuits cannot produce duty cycles below 50%. The circuit exploits the symmetry of the trigger and threshold inputs by swapping the positions of C1 and R1. The duty cycle is adjustable to up to approximately 75% with P1.

resistors, and capacitors, interconnected where needed with bits of gold wire. Noyce, on the other hand, used silicon and connected the parts with copper lines directly "printed" on the crystal using Hoerni's planar process. Noyce's device is considered world's first monolithic IC. It was suitable for mass production, something not possible with Kilby's design.

As mentioned above, Kilby's circuit contained resistors and capacitors besides transistors. Resistors are quite easy to create in an IC, but capacitors and inductors are more problematic as even small values already require a lot of chip space ('real estate'). Even though techniques exist to simulate reactive components with transistors, this is the reason why many ICs require the connection of external capacitors and inductors to make them work.

Today, nearly all modern ICs are based on MOSFETs (Metal-Oxide-Silicon Field-Effect Transistors) instead of bipolar junction transistors (BJTs) because MOSFETs require fewer production steps in IC manufacturing.

As of 2025, world's best-selling integrated circuit is still the NE555 timer IC designed by Swiss electronics engineer Hans Rudolph Camenzind (1934–2012)[17]. It has held this position almost since its introduction by Signetics in 1972. Of course, heaped together, op-amps and microcontrollers are used more, but that's when you count every reference from every manufacturer. The NE555, or simply '555', is the most-used single-reference IC. (It is also the first IC I have worked with myself.) The 555 can be found in applications varying from timers and oscillators to power supplies, audio amplifiers, and more. One of the reasons for its popularity, besides the fact that it is a very practical and low-cost device, is that Signetics did not patent

17 Hans Camenzind's YouTube channel is still online (in 2025). It has a few videos about famous inventors and inventions, and an analysis of freeway traffic jams. https://www.youtube.com/@hcamen

the 555. As a result, it was quickly copied by competitors world-wide, and several clones appeared within a year after its introduction. According to (a surprised) Camenzind, and as far as we can tell, no one ever went through the trouble of trying to improve the design of the 555, even though Camenzind published a list of modifications that would make it a better device. The 556 combines two 555s in a single IC and is still an active product. The 558 integrated four 555s but to make it all fit in a standard 16-pin package, the 555s were simplified. Probably due to this, the 558 was not a big commercial success.

Fun with Algebra

The name '555' was chosen arbitrarily by the marketing manager of Signetics, the company that released the IC. The sum of the digits of 555 is 15. If you divide 555 by 15, you obtain 37. Therefore, 555 is a Harshad number, an integer that is divisible by the sum of its digits (just like 666, the number of the beast). Furthermore, 555 is a palindrome in base 10, and it is a repdigit.

Initially, the integrated circuit was monolithic, meaning that all the circuitry was built on a single die (crystal). Actually, that was the whole point of it. Even today, most ICs are made this way. However, miniaturization and manufacturing techniques have advanced a lot since the 1960s, and applications have become increasingly demanding. In 1975, American engineer Gordon Moore (1929–2023) noted that the number of transistors in an integrated circuit doubled approximately every two years. This observation has become an empirical law that still holds true more than 40 years later. Even though transistors have almost shrunken to the size of an atom and can hardly get any smaller, semiconductor manufacturers manage to cram more of them in a single IC, keeping up with Moore's law. They do this by stacking several dies (as many as sixteen or more) in a single IC package. The result is a three-dimensional IC (3D IC). Monolithic, i.e. lithographically stacked 3D ICs are being developed too but it is, as you can imagine, a complicated process. Another neat trick to get more transistors in the same board space is by mounting two ICs on top of each other using package-on-package (PoP) technologies. PoP tries to combine the benefits of die stacking and traditional IC packaging while minimizing their limitations.

Feedback

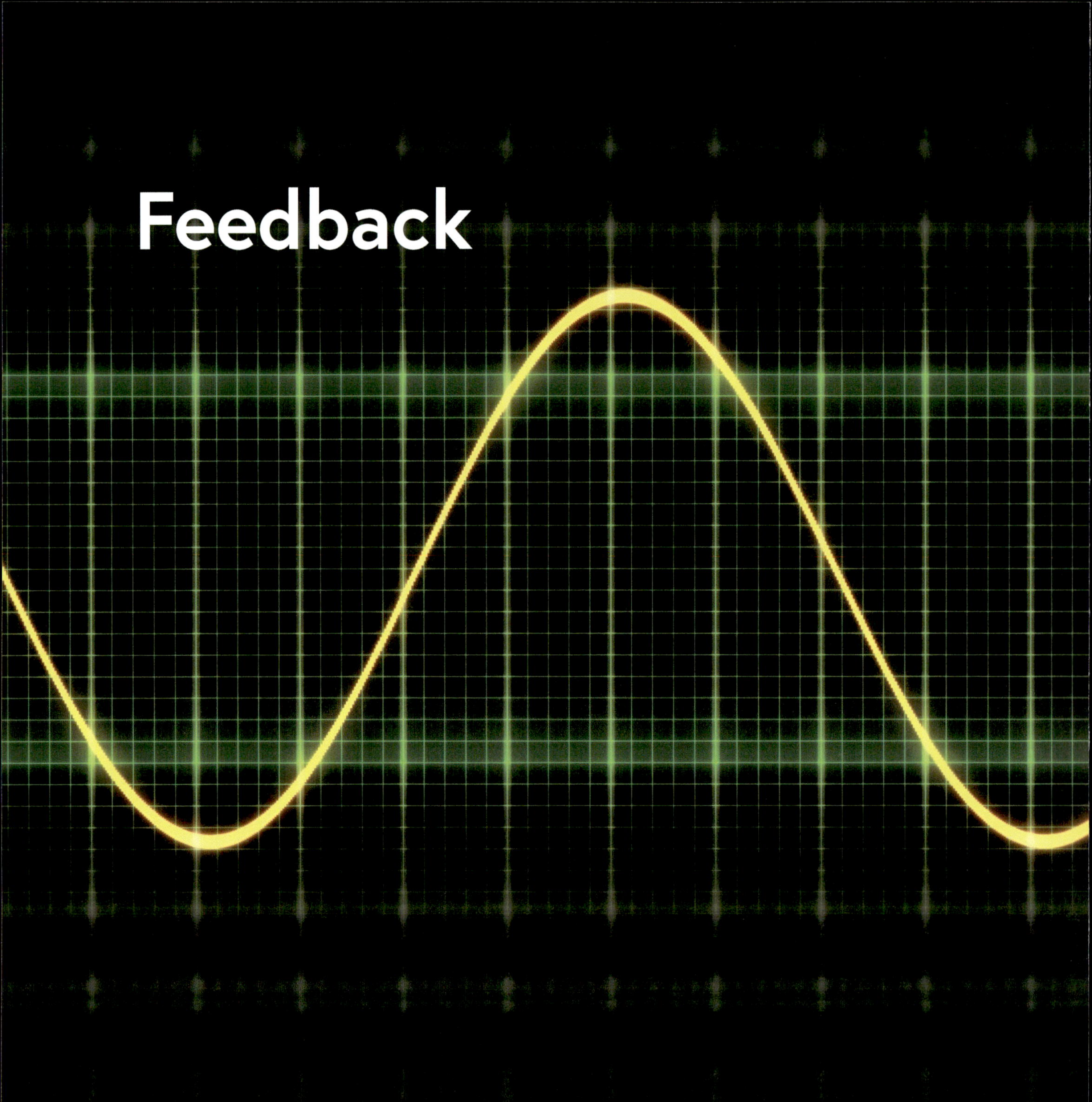

On August 8, 1928, American electrical engineer Harold Stephen Black (1898–1983) filed a complicated patent for a 'Wave Translation System'. The first claim reads as follows:

1. In a wave translating device or system having amplifying properties, an input portion and an output portion, means to apply fundamental waves to said input portion, said system carrying fundamental components in said output portion and having means producing other wave components in said output portion, and means controlling the relative magnitudes of said components in said output portion comprising means to feed waves from said output portion to said input portion to decrease the gain of the system.

Trying to digest this and knowing that there were 126 similarly worded claims in total, you can easily understand why it took the patent office almost ten years before the patent was granted (on December 21, 1937. According to Black, it was because they didn't believe that it worked.)

The takeaway from Black's patent is the concept of negative feedback applied to an amplifier, a revolutionary idea that changed electronics forever. Nobody likes negative feedback, we humans prefer constructive criticism, but when listened to, it can help make things better. And that's what it does to an amplifier. A mediocre amplifier without feedback can be turned into a great amplifier by applying negative feedback. (Maybe we should call it a constructive criticism amplifier?)

Missed Opportunity

As is often the case with inventions and discoveries, several people can come up with similar ideas. English audio engineer Paul Gustavus Adolphus Helmuth Voigt (1901-1981) filed a patent in 1924 describing a feedback amplifier. However, he did not provide the theoretical background as did Black

So, what is feedback? It is simply feeding some of the output of a process back to its input. It can be negative, meaning that the feedback signal opposes the input, or it can be positive so that the feedback signal will encourage the input. In dog-education terms, you apply positive feedback when

you want your dog to learn a new trick by giving it snacks and saying how good a boy or girl it is. Negative feedback is applied when you want your dog to stop attacking the pizza delivery person. I am sure you have an animal-friendly method for doing that.

The opposite of feedback is feedforward, a technique that is also used in electronics. As a matter of fact, that's what Black tried first, before turning to negative feedback. Feedforward systems compare the output of a process to the desired result and then try to shape the output to make it match the desired result as closely as possible. It's what you do when you cut off the burnt parts of your steak. It's a kind of symptomatic or palliative treatment as it doesn't try to fix the cause of the problem (by insulting the cook). Feedforward correction techniques are useful when the root cause cannot be changed because it is a natural phenomenon, or because you don't really understand what's going on, or when it is not worth the trouble (i.e., the money).

Harold Black was working on complex telephone systems that included many amplifiers (repeaters) connected in series. Each amplifier added a bit of noise and distortion to the signal going through them. As the amplifiers weren't exceptional, it was very complicated to obtain an acceptable sound quality at the end of the line. He had tried feedforward correction but wasn't satisfied with the results. Then, as it often happens with inventions, he had a 'Eureka!' moment and negative feedback was born.

Negative feedback applied to an amplifier is a bit counterintuitive as you pay for it with amplification. In order to improve the amplifier, you take away some of its gain. As a result, the amplifier doesn't amplify that well anymore (but boy, does it sound good!). Because of this, Black had a tough time convincing his pears. Black's solution was to provide the amplifiers with more gain than necessary and then use the extra gain to compensate for the negative feedback.

The drawing shows an amplifier with negative feedback applied to it. The open-loop gain, i.e., when there is no feedback at all, is represented by *A*. Looking at this drawing, we can write down the following two equations:

$$V_o = A \cdot V_i^*$$

$$V_i^* = V_i - \beta \cdot V_o$$

Combining the two into one by substitution of V_i* gives:

$$V_o = A \cdot (V_i - \beta \cdot V_o)$$

Rearranging this, we obtain:

$$(1 + \beta \cdot A) \cdot V_o = A \cdot V_i$$

Rearranging once more, we can define the feedback gain A_{FB} like this:

$$A_{FB} = \frac{V_o}{V_i} = \frac{A}{1 + \beta \cdot A}$$

When β is zero, i.e., no feedback, then A_{FB} is simply equal to A. When A is large, which is usually the case, then we can simplify the equation by dividing by A:

$$A_{FB} = \frac{A}{1 + \beta \cdot A} = \frac{1}{\frac{1}{A} + \beta} \approx \frac{1}{\beta}$$

This shows two things: first, more feedback implies less gain, and second, the open-loop gain A has no influence when it is large. Therefore, by incorporating negative feedback, the amplifier design becomes less dependent on component value tolerances and mismatches. This very important result is one of the reasons why negative feedback is so widely applied today.

The feedback-gain equation also shows that when $\beta \cdot A$ is close or equal to −1, the system becomes unstable and may start to oscillate, or the output might grow uncontrollably (saturate) in one direction (runaway). The theory describing this had already been worked out in 1932 by Swedish physicist Harry Nyquist (1889–1976) while Black was waiting on his patent to be granted. Nyquist not only provided the theory, but he also formulated rules for ensuring stability in feedback systems. This became known as the Nyquist stability criterion[18]. American engineer Hendrik Wade Bode (1905–1982) imagined an easier but less general graphical method, now known as Bode plots. These are graphs of the

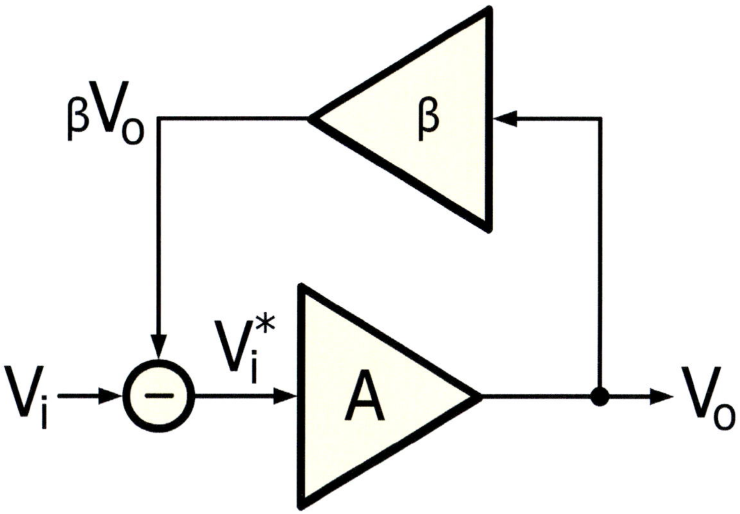

An amplifier with negative feedback. The open-loop gain, i.e. the gain without feedback, is *A*.

18 Independently discovered in 1930 by German electronics engineer Felix Strecker (1892–1951).

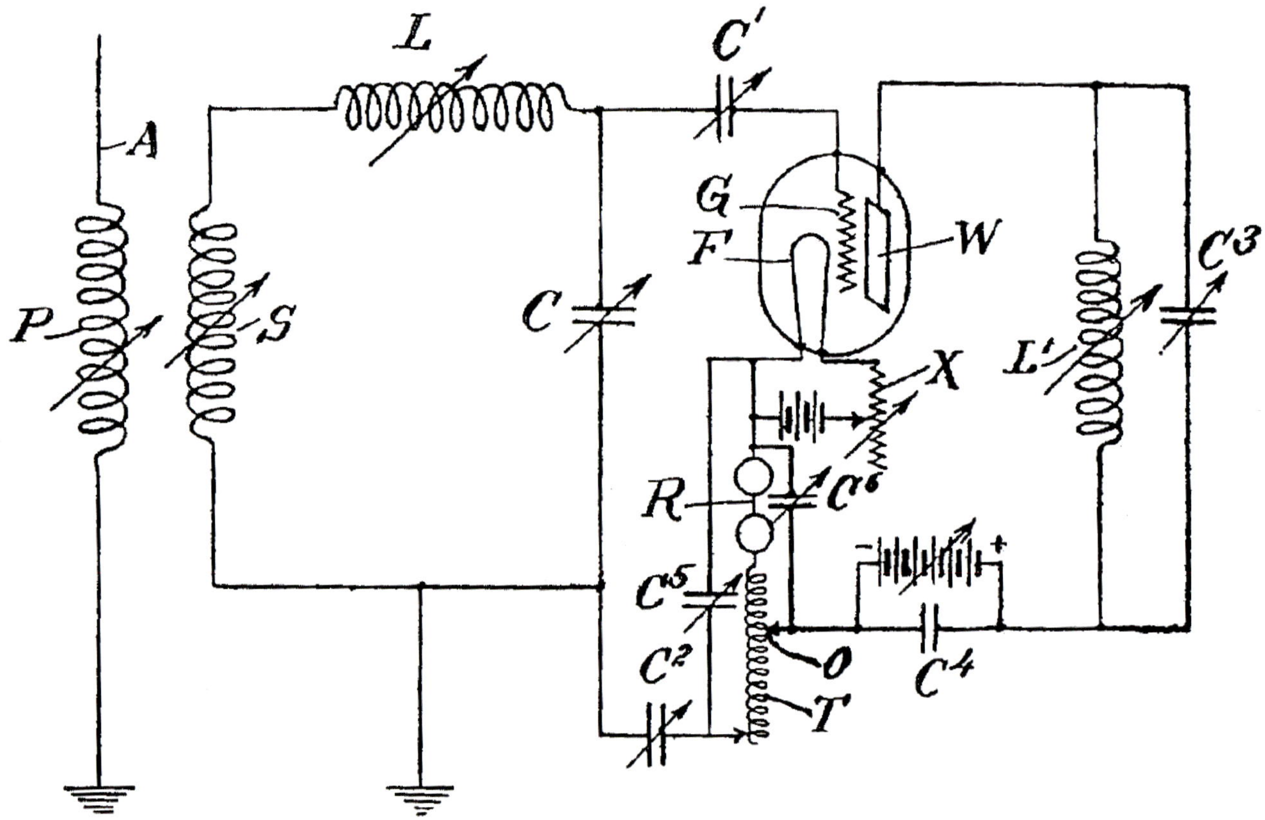

In 1914, Armstrong filed a patent for a wireless receiving system based on positive feedback and that became known as the 'regenerative circuit'. It was attacked afterwards in 1915 by Lee de Forest, which started yet another endless patent dispute.

frequency response of a system showing its magnitude and phase responses. Both Bode and Nyquist plots have become important tools in application domains like control theory and filter design.

In electronics, positive feedback exists too and is, like negative feedback, widely used. American electrical engineer Edwin Howard Armstrong (1889–1976) is considered the first to have applied positive feedback. In 1912, he discovered that by routing part of the output signal of an amplifier back to its input, the gain of the amplifier increased. This allowed weak radio signals to be amplified enough to be detected, an important step forward in radio technology. This may sound somewhat contradictory after having just read how negative feedback improves an amplifier by reducing its gain, but it is easy to understand when looking at the signals that are being amplified.

Black worked with relatively strong audio signals that occupy a range of frequencies (300 Hz to 3.4 kHz in the case of telephony), whereas Armstrong tried to amplify a weak amplitude-modulated signal with a single frequency (the carrier). Black wanted to amplify all the frequencies in his signal in a consistent way to maintain good audio quality. Armstrong, on the other hand, wanted to recover only the (varying) amplitude of his carrier signal by means of detection. Therefore, the high amplification of a single frequency provided by the positive feedback was perfect for him. Armstrong's circuit, patented in 1914, became known as the 'regenerative circuit' because it regenerates the input signal by feeding back a portion of the amplified output signal in phase with the input

Armstrong pushed the concept of positive feedback further by making the amplifier almost oscillate to obtain even more gain. By interrupting the feedback signal periodically—'quenching' it—the amplifier is reset before its starts to oscillate. This technique is called super regeneration.

Positive feedback may lead to oscillation, and this is another useful application of the technique. Oscillators are used in many applications and form an important subject of research in electronics. There are other ways to create oscillators, like the relaxation oscillator, but positive feedback works well too. A common example that probably everyone knows is the Larsen effect (a.k.a. audio or acoustic feedback, or simply feedback) that occurs when a microphone comes too close to a loudspeaker in a sound system. The squealing sound that results is due to positive feedback.

In 1921, German physicist Heinrich Georg Barkhausen (1881–1956) derived the first mathematical conditions for oscillation in electrical circuits, a kind of opposite of Nyquist's stability criterion used to avoid oscillation. The so-called Barkhausen stability criterion specifies the necessary conditions for sustained oscillation. Referring back to the feedback gain equation presented earlier

$$A_{FB} = \frac{A}{1 + \beta \cdot A}$$

Armstrong also invented the supersonic heterodyne receiver (a.k.a. super heterodyne or 'superhet') and FM radio.

Oscillators never start; amplifiers always oscillate...

Barkhausen's stability criterion states:

1. The absolute magnitude of the loop gain must be equal to one, i.e. $|\beta \cdot A| = 1$.

2. The phase shift around the loop must be zero degrees (or an integer multiple of 360°).

Remember, stability in this context refers to sustained oscillation.

Note that Barkhausen's criterion is a necessary condition for oscillation, but not a sufficient condition. Electronics engineers know this from experience, as is conveyed by the well-known adage: "oscillators never start; amplifiers always oscillate". Similarly, when an oscillator is created in a circuit simulator program on a computer, chances are that it will not work even though the same circuit built with real components on a breadboard works fine. To make the simulation work, it is often required to introduce a non-linearity or noise source. Does this indicate that Barkhousen's rule should be extended with a third criterion?

3. The circuit must be non-linear up to a certain extent.

The question appears to be still open.

A well-designed oscillator is an example of a bootstrap process, a self-starting process that continues without external input. Bootstrapping refers to the saying "to pull oneself up by one's bootstraps". Even though oscillators rely on bootstrapping to start, in electronics the term usually refers to the use of positive feedback to increase the input impedance of an amplifier. Bootstrapping can help a weak signal. Normally, the input signal uses ground for reference and the impedance seen by the signal is the impedance to ground.

According to Ohm's law, the current flowing through the input impedance depends on the signal's voltage level. When it is close to ground, a small current will flow, and when the level is high, the current will be larger. When bootstrapping, a portion of the output signal is fed back in phase to a point near the input where it is used as the reference instead of ground. The difference between the signal and its copy is very small, and therefore only a tiny current will flow through the impedance between the two. The input impedance has become very high. The feedback signal provides the rest of the current that would be needed without feedback. The weak signal is placed on the shoulders of the strong feedback signal.

Bootstrapping

It is often said that the term bootstrapping refers to Baron Munchausen (1785) who pulled himself and the horse he was sitting on out of a swamp by pulling his bootstraps. This is not true, however. As the lithograph by Swiss painter Martin Disteli (1802–1844) shows, Munchausen did not have bootstraps and instead pulled his hair to deliver himself from this embarrassing position. Like bootstrapping, hair-pulling too is a common technique used in electronics, especially when trying to solve frustrating problems.

The Operational Amplifier

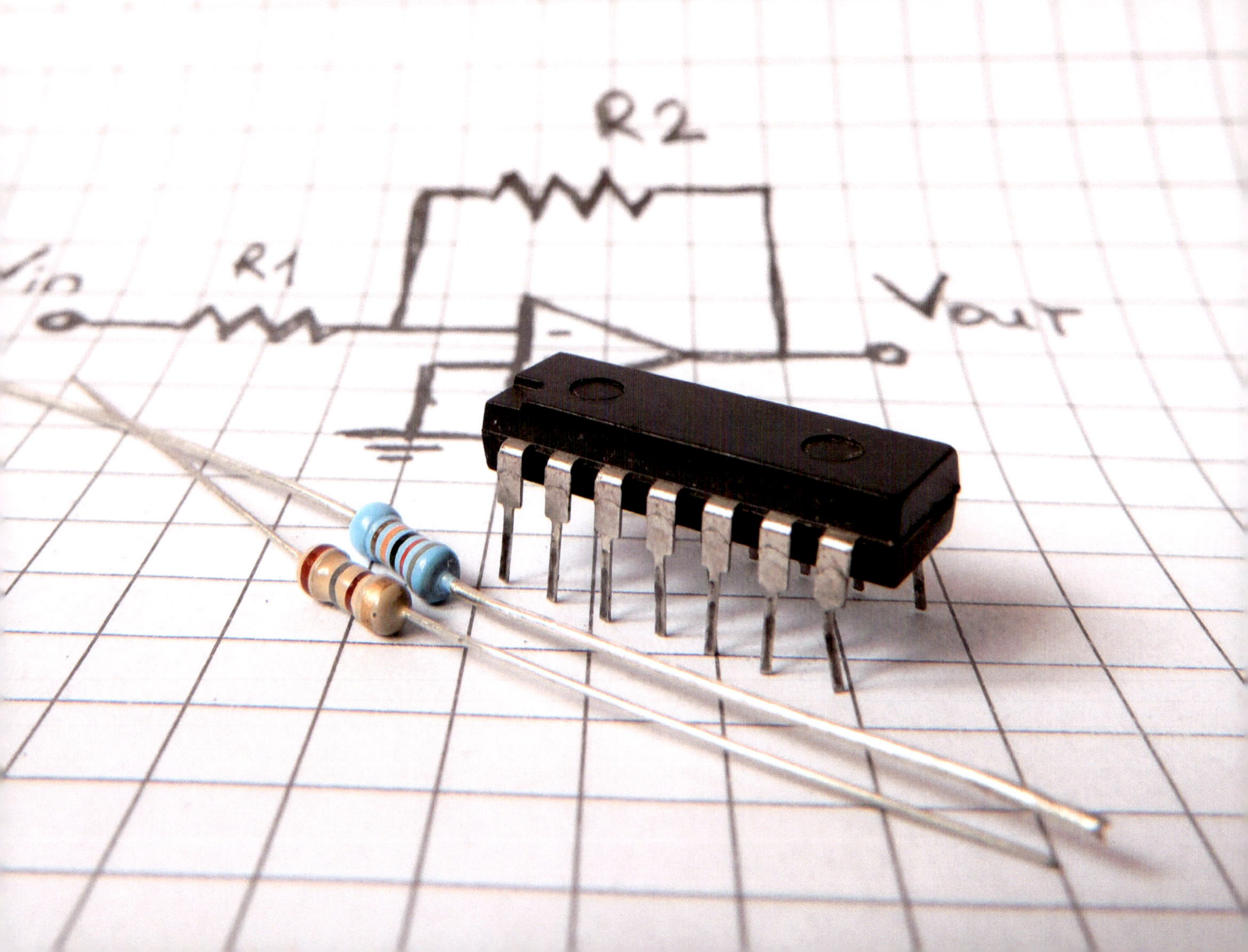

Most components used in electronic circuits are relatively simple devices. The resistor is basically just a piece of carbon, an inductor is a bit of metal wire wound as a spiral, and the capacitor consists of two metal plates placed parallel to each other. Semiconductors like diodes and transistors are not very complex either as they are simply layers of p- and n-type silicon stacked on top of each other. Basically, electronics is nothing more than mixing burned wood, sand and copper. The Neanderthals could have come up with it if they hadn't been so focussed on flint. As a matter of fact, flint is mostly silicon dioxide (SiO_2), so, in a way, we might say that prehistoric people manipulated semiconductors, not to make transistors but to make sparks. Interestingly, early wireless transmissions were based on sparks too, but I digress.

Besides the components mentioned above, a compound device has become very important too in electronics: the operational amplifier or op-amp. The op-amp is a relatively small circuit made from the basic electronic components, but in the shape of an integrated circuit. It has two inputs, one inverting ('–') and one noninverting ('+'), and one output. The op-amp therefore is a kind of triode, a three-terminal device.

The properties of the op-amp are as well defined as those of the discrete passive (R, L & C) and active (diode & transistor) components. A single equation, only slightly more complex than Ohm's law, describes its behaviour:

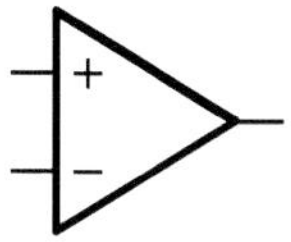

The electric symbol of the op-amp.

Here, as shown in the drawing, Vo is the output voltage, V+ is the voltage on the noninverting input, V– is the voltage on the inverting input, and A is the so-called open-loop gain (amplification) of the op-amp. The superpower of the op-amp lies in the amplification, which is infinite for the ideal case. The ideal op-amp also has infinite input impedance, zero output impedance, and infinite bandwidth, i.e. it can amplify signals of any frequency. However, before we take a closer look at how a device featuring infinite amplification can be useful, lets investigate its origins.

In the 1930s, electronic computers were used a lot for military applications like plane simulation and automatically aimed anti-aircraft guns. Known today as analogue computers, they contained ampli-

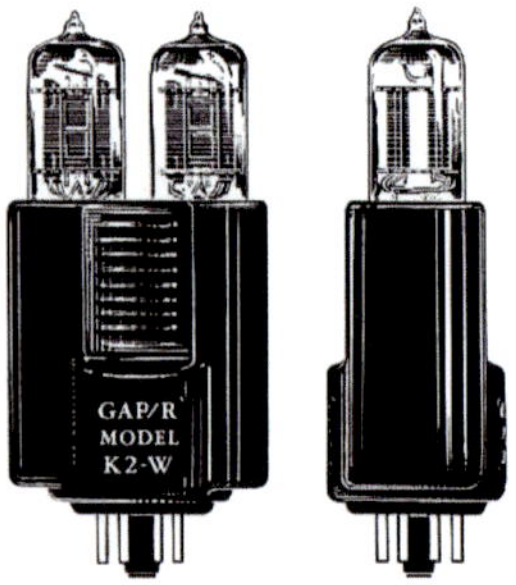

The GAP/R Model K2-W was the first commercially available op-amp.

fier modules that performed the required mathematical operations. American electrical engineer John Ralph Ragazzini (1912–1988) proposed the term 'operational amplifier' to refer to these amplifiers. (He wasn't happy with it, but it stuck.) In a paper from 1947, he defines it as follows:

"The term 'operational amplifier' is a generic term applied to amplifiers whose gain functions are such as to enable them to perform certain useful operations such as summation, integration, differentiation, or a combination of such operations."

Ragazzini did not invent the op-amp, that had been done a few years earlier by Karl Dale Swartzel Jr. (1907–1998) who filed a patent for it in 1941. In it, he describes a summing amplifier that uses feedback to create high-impedance inputs and a well-defined amplification. As we will see later, this is exactly how op-amps are being used today. Swartzel's device had only one input, the inverting one, as the non-inverting input was connected internally. Swartzel's op-amp used, of course, tubes because the transistor hadn't been invented yet.

The op-amp as we know it, i.e., with differential inputs, was invented in 1942 by American engineer Loebe Julie (1920–2015). His design, intended for radar-guided anti-aircraft guns, introduced the differential inputs '+' and '–' mentioned at the beginning of this chapter. It was smaller, simpler, much faster and consumed less power than the designs in use at the time. American engineer and entrepreneur George Arthur Philbrick (1913–1974) took Julie's design and turned it into the first commercially available op-amp, the GAP/R Model K2-W. It was launched in 1952 as

"a general-purpose high-gain d-c amplifier of modular construction, offering stability and fidelity for operational service in feedback computing systems of every speed."

Philbrick's K2-W used two tubes and remained available until circa 1974, even though the monolithic-semiconductor-based IC op-amp had been around for at least a decade. Indeed, introduced in 1964, the μA702, designed by American electronics engineer Robert John Widlar (1937–1991), was the first monolithic operational amplifier IC. It was followed a year later by the μA709 (again by Widlar), which became an instant hit. Initially, the device was only available in a round metal can with eight leads, but an 8-pin dual-in-line (DIL) package version was released soon afterwards. This packaging and its pinout became the industry standard for op-amps

The Discrete Operational Amplifier

In the period between the tube-based op-amp and the monolithic-IC op-amp, a third species of op-amps existed briefly: the discrete semiconductor op-amp[19]. These were operational amplifier modules built with discrete components like resistors and transistors. The Philbrick P45 was such a module, designed in 1963 by American electronics engineer Robert Allen Pease (1940–2011). The Philbrick company GAP-R did a lot to turn op-amps into easy-to-use standardized modules, which contributed a lot to the op-amp-as-a-component concept. Later, such modules became hybrid op-amps, ICs that instead of a silicon crystal had a tiny circuit board inside.

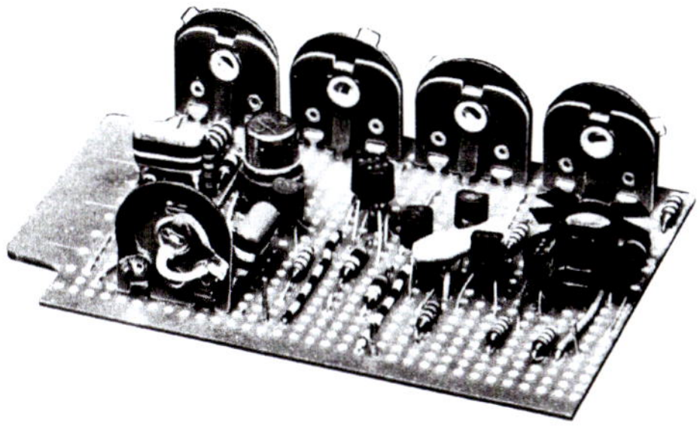

The operational amplifier shown here, published in the October 1966 edition of the Dutch DIY electronics magazine Elektuur, had a lot of adjustment points and was intended for prototyping. It featured an open-loop gain of 500,000, an open-loop bandwidth of 10 kHz, a closed-loop bandwidth of 1 MHz, an input impedance 20 MΩ, an output impedance of 25 Ω. These were better specifications than those of the μA709 that was priced at around $300 in 1966. The DIY op-amp costed less than $30 (which corresponds to almost $300 in 2025). As of 2025, general-purpose op-amps can be had for as little as 10 cents.

and made the device even more popular. However, it was the μA741 launched in 1968 that became the father of all future general-purpose operational amplifiers. The μA741 was designed by English electronics engineer David Fullagar (1942–) at Fairchild Semiconductor to counter National Semiconductor's LM101 (designed by Widlar, who had changed sides). More reliable and easier to use than previous op-amps thanks to its built-in compensation capacitor, the μA741 quickly found its way into many thousands of products. At the time of writing, the '741' is still being produced by major semiconductor manufacturers even though hundreds of different, higher performance alternatives are available.

19 Discrete op-amps are still being used in applications with particular specifications.

Since the introduction of the first op-amp, the number of different models has steadily increased. Each new op-amp performed better in a certain way. Faster, less offset at the output, higher-impedance inputs, lower noise figures, wider output range, improved performance of whatever parameter you can think of. Today, there are so many op-amps around that they are usually classified by specialization, like general-purpose, audio, video, precision, low-voltage, transimpedance, etc. Choosing the best op-amp for an application has become a time-consuming and complicated task.

Now, let us have a brief look at how the operational amplifier with its infinite gain can be useful. Remember the equation at the beginning of this chapter.

$$V_o = A \cdot (V_+ - V_-)$$

Dividing both sides of the equal sign by A, we get

$$\frac{V_o}{A} = V_+ - V_-$$

Because the value of A is very large (infinite for an ideal op-amp), this equation can be simplified to:

$$V_+ = V_-$$

This is an important result to keep in mind when analysing op-amp circuits. In the presence of feedback, this implies that an op-amp will adjust its output voltage in an attempt to make the voltages on its inputs the same. The only way to achieve this is by providing a path from the output of the op-amp to at least one of its inputs. Almost every op-amp circuit has some form of feedback. Let's see how this works in practice.

The inverting amplifier as shown here is a very common op-amp configuration. Understanding how it works only requires the laws of Ohm and Kirchhoff.

As explained above, since R_2 connects the output to the inverting input, it provides feedback. Therefore, assuming that the op-amp is an ideal device, the voltages on the op-amp's inputs may be considered equal. Thus, if we assume that $V_+ = 0$ V (Why? Why not.), then V_- is also equal to 0 V. The op-amp being ideal, the impedance of its inputs is infinite. This implies that no current can flow into them. Applying the laws of Ohm and Kirchhoff, we may therefore write:

$$\frac{V_o - V_-}{R_2} = -\frac{V_i - V_-}{R_1}$$

Kirchhoff's current law that states that the total current entering a junction (here V_-)

is equal to the total current leaving the junction. Therefore, the current through *R2* must be opposite to the current through *R1*, hence the minus sign.

Since V_- is equal to V_+, which we decided was zero, it can be removed from the equation:

$$\frac{V_o}{R_2} = -\frac{V_i}{R_1}$$

Shuffle it around a bit (i.e., multiply both sides by R2) to obtain:

$$V_o = -\frac{R_2}{R_1} \cdot V_i$$

This equation states that the output voltage is equal to the input voltage multiplied by a factor of –R2/R1. In other words, we have created an amplifier with a gain of –R2/R1. The minus sign shows that it is an inverting amplifier.

The analysis technique presented above can be applied to many op-amp circuits that provide a path from the op-amp's output to (one of) its inputs, like the noninverting amplifier, the integrator, the differentiator, the active filter, and so on. But..., what if there is no such path? Then it is easy too. Again, it starts by looking at the first equation of this chapter

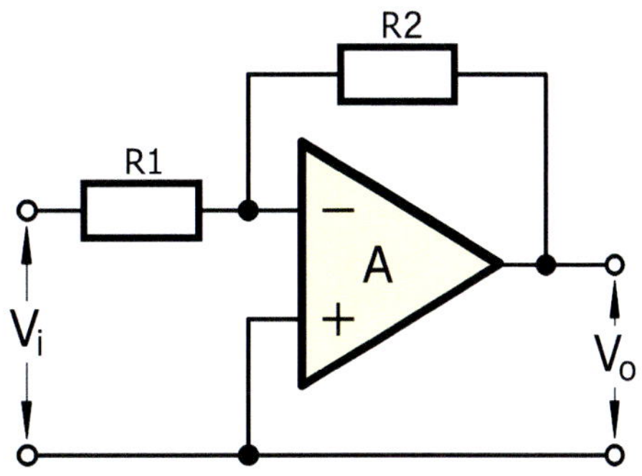

The op-amp as inverting amplifier. The two resistors set the voltage gain of the amplifier. *R2* is used as a negative feedback resistor. The noninverting input is connected to the common reference (usually called ground) and the voltage on it can be considered to be zero.

$$V_o = A \cdot (V_+ - V_-)$$

It says that the output voltage is the difference of its two input voltages multiplied by A. Since *A* is very large (infinite, in theory), even a tiny difference between the two input voltages will produce either a very large positive value (when V_+ is larger than V_-) or a very large negative value (when V_+ is smaller than V_-) at the output[20]. The exact value is only limited by the supply voltage and the internal design of the op-amp. Is this behaviour useful? Yes, it is, because it allows you to use the op-amp as a voltage comparator to make decisions based on voltage levels. `If` V_+ is larger than V_- `then` do this; `else` do that. Analogue computing, remember?

20 When V_+ and V_- are exactly equal, V_O should be zero. In practice, however, this is almost impossible to achieve because the two inputs are not perfectly identical due to component tolerances.

Optoelectronics

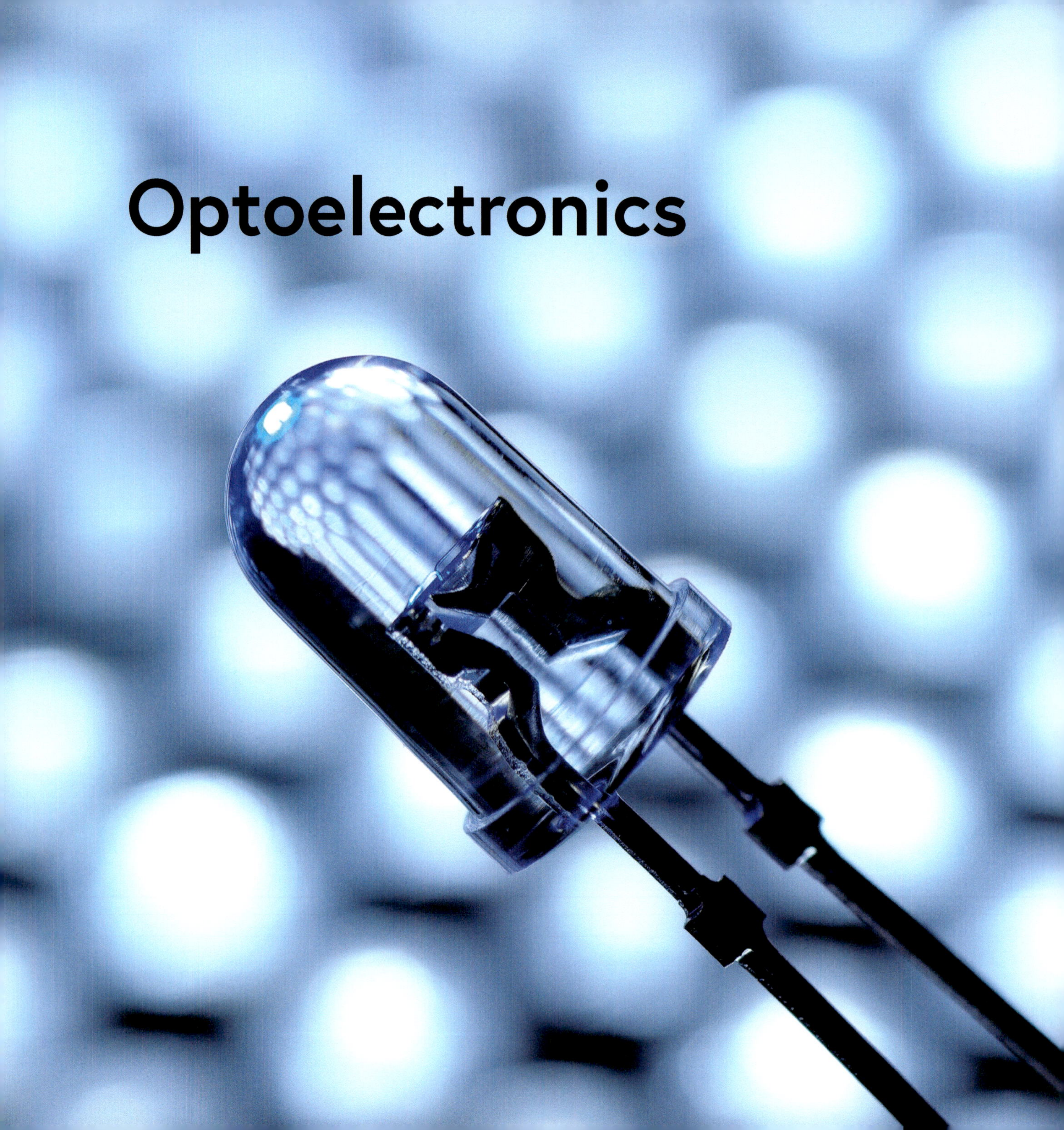

For more than a century, we have heated our houses with inefficient incandescent lightbulbs that produced warm light, literally. Today, we must heat our homes in another way as the incandescent lightbulbs have been replaced by efficient LEDs that spread cold light around.

The LED, the light-emitting diode, is an optoelectronic device, a device that combines light and electronics, but optoelectronics is not limited to LEDs. Optoelectronics existed before the invention of the LED. The influence of light on electronic properties of materials has been known since the first half of the 19th century. Hertz observed the photoelectric effect in 1887 but didn't try to explain it. It was the German-Swiss theoretical physicist Albert Einstein (1879–1955) who, in 1905, came up with a plausible explanation of how light can induce an electric current. Einstein introduced the concept of light as discrete packets of energy (light quanta) instead of as waves as predicted by Maxwell's equations. In 1926, Einstein's light quanta became photons. Russell Ohl discovered in 1939 that the p-n junction of a semiconductor material produces a small current when exposed to light by absorbing photons and invented the solar cell in 1941. The phototransistor, a light-sensitive transistor, was invented in 1948 by the American physicist John Northrop Shive (1913–1984).

Top Nobel Prize Nominee

The term 'photon' was coined in 1926 by American physical chemist Gilbert Newton Lewis (1875–1946). It is often said that Lewis was nominated 41 times for the Nobel Prize in Chemistry but never won. However, this is not as unfortunate as it may seem, as 41 refers to the number of people who nominated him. Between 1922 and 1946, he was officially nominated only 17 times—almost every year until his passing.

In 1907, English engineer Henry Joseph Round (1881–1966), while experimenting with cat's whisker detectors, observed light emissions from his crystals. He described yellowish electroluminescence in a silicon carbide (carborundum) crystal. Around 1924, Russian scientist Oleg Vladimirovich Losev (1903–1942) had noticed the same

Light-Emitting vs. Laser

A light-emitting diode or LED produces incoherent non-directional light that spreads in many directions. A laser diode on the other hand emits highly directional and focused coherent light. This difference is due to the mechanisms behind the two phenomena. The LED is based on spontaneous emission, where electrons emit photons when they recombine with holes. The result is diffuse, unfocussed light in a narrow band around a central wavelength. The opposite of spontaneous emission is stimulated emission where photons stimulate other electrons to emit identical photons. As a result, the emitted light is of a single wavelength (monochromatic) and is focussed as a narrow beam. This process is used by laser diodes. LED light is intended to be seen and admired; direct laser light should not enter your eye.

effect and turned it into a practical device, the first LED (1927). However, as it was not very efficient, it was quickly forgotten. It was reinvented as an infrared (IR) light source in 1962 by American engineers Gary Pittman (1930–2013) and James Robert Biard (1931–2022) . However, American engineer Nick Holonyak Jr. (1928–2022) became the father of the LED after he demonstrated in 1962 the first practical device that emitted visible red light. He also invented the first visible-light laser diode.

Henry Round observed the emission of photons from a semiconductor material, while Russell Ohl noticed their absorption. Indeed, optoelectronics encompasses both phenomena, photon emission and absorption. An interesting electronic device that illustrates this nicely is the optocoupler or photocoupler. It consists of an LED and a phototransistor together in a single package. The input is the LED, the output is the transistor. When the LED shines on the transistor, the transistor turns on. Otherwise, the transistor is off. Usually, it is used as a two-state switching device, but gradual operation is sometimes possible too. In case you wonder why one would need a device that turns electric current into light and then back into electric current, its main use is to isolate two circuits from each other. A photocoupler provides galvanic isolation between two circuits. Therefore, it is also referred to as opto-isolator. Galvanic refers to direct current (DC). It doesn't allow DC to flow between two circuits, but AC can flow as for instance current, light, or magnetic fields. A capacitor is a galvanic isolator and so is a transformer.

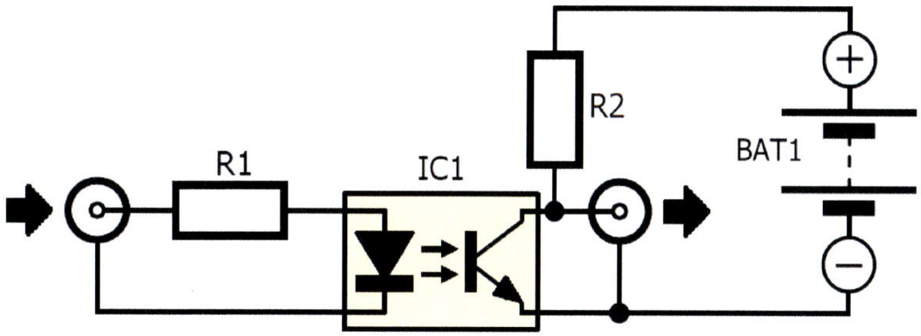

The opto- or photocoupler IC1 combines an LED and a phototransistor in a single package. The electric current that flows into the input generates an electric current at the output even though there is no electrical connection between the two circuits. Note that the above circuit act as an inverter. When the LED is on, the transistor conducts and connects the output to minus. When the LED is off, the transistor blocks and the output is connected to plus (through R2).

The light detector doesn't have to be a phototransistor or photodiode, light-dependent resistors (LDRs) are used too. Today, such devices have become hard to find, but you can make one yourself with a light (LED or lamp), an LDR and some black (opaque) self-adhesive tape. Such devices can replace a resistor in a circuit and make the circuit light sensitive. This allows for easy control of almost any parameter of a circuit, like the gain of an amplifier or the cutoff frequency of a filter. Such contraptions are also suitable for adding feedback to a circuit to stabilize it.

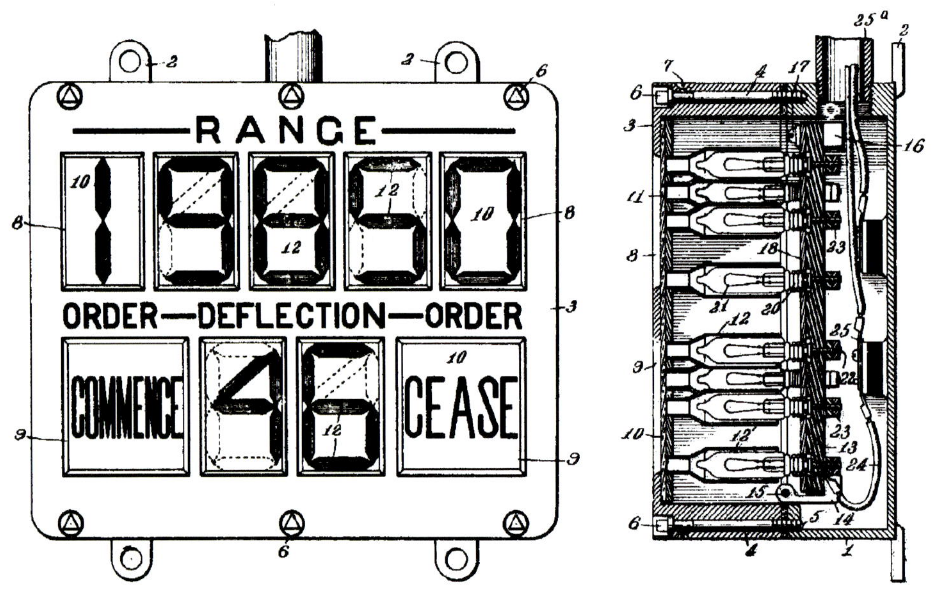

The 7-Segment Display

The 7-segment display used in millions of clocks and other appliances and instruments is an emblematic application of LEDs. But did you know that the idea for it was invented many years before the creation of the LED? It was a certain Frank W. Wood who, in 1908, filed a patent for an "illuminated announcement and display signal", a large display device intended for warships and the like. To keep maintenance of the display to a minimum, it used standard incandescent light bulbs. To allow displaying any digit with as few lamps as possible, Wood used a segmented character set for digits[20]. The main difference between Wood's character set and the modern 7-segment display is that Wood used eight segments instead of seven. The eighth segment was used for the diagonal of the digit '4'. Later, the eighth segment was turned into a decimal point and the '4' became an 'h' rotated by 180 degrees. Woods segments were vertical and horizontal where modern 7-segment digits typically feature a slant of approximately 10 degrees. Displays with more segments exist too, like nine, fourteen or sixteen. These can display more characters and support non-western character sets.

21 Frank Wood probably did not invent the segmented character set but adapted it.

When the LED (or another light source) and phototransistor (or another light detector) are mechanically separated from each other, one obtains a light barrier. Such barriers are used a lot in automation to detect objects or persons. A system equipped with optical barriers can then decide to open or close a door, stop or start a conveyer belt, or take some other action. Light barriers can span anything from millimetres to thousands of kilometres. The distance between the Earth and the Moon for instance is measured with a laser bouncing of a reflector placed on the Moon. In this case, the distance between the light source (laser) and light detector is roughly 750,000 kilometres.

Separating the light source and the detector also allows for wireless communication or (long-distance) communication over fibre. As an example, the remote control of your TV set uses infrared light to transfer your commands to the TV, while your internet connection may use optical fibre. Because LEDs and lasers can be switched on and off superfast, the modulated light they produce can convey large amounts of information. Applications range from infrared wireless headphones to satellite communication. In 2023, the NASA's TeraByte InfraRed Delivery (TBIRD) system achieved a 200 Gbit/s space-to-ground laser-communication link with a satellite orbiting at 530 km. In 2024, as part of the Deep Space Optical Communications (DSOC) program, NASA established a 6 Mbit/s link with a spacecraft at a distance of more than 450 million kilometres from Earth. That's a faster internet connection than most of us have at home.

Optical computing may come to mind when thinking about optoelectronics. Optical computers exploit not only the high data rates possible with light, but also some of light's special properties. Optoelectronic devices only provide the interface between the optical computer and the electronic circuits around it. Optical computing itself is outside the scope of this book.

Believe it or not, but you can win the Nobel prize simply by inventing a little blue light. Japanese engineer and scientist Shuji Nakamura (1954–) achieved this feat in 2014 after he had invented the blue LED in 1993. He shared the prize with Japanese scientists Isamu Akasaki (1929–2021) and Hiroshi Amano (1960–). Besides the Nobel prize, they have received several other prestigious international awards for their invention. Why? Because the blue LED was key to create the holy grail of LEDs, the white LED. The latter forms the basis of solid-state lighting as we know it today

In the human eye, yellow stimulates both the red and green photoreceptor cells, while blue takes care of the blue photoreceptors. Mixed together, the brain perceives it as white light.

and has caused a revolution since its introduction. White light is obtained by passing blue light through a yellow phosphor coating. A part of the blue light is converted to yellow, and the human eye does the rest by mixing yellow and blue into white. By controlling the mix of blue and yellow, the tone or temperature of the light (in Kelvin) can be adjusted.

The blue LED was also required for full-colour displays. Thanks to Nakamura's invention, today we all enjoy lightweight, LED-based flat-screen television sets and full-colour displays for smartphones, smartwatches, and tablet- and laptop computers. In a full-colour display, every pixel consists of a red, green and blue LED. By mixing the intensities of each LED, in theory, any colour can be obtained. A television

set with a 4K resolution has 8,294,400 pixels (3840 × 2160). This corresponds to almost 25 million LEDs (including the white backlight LEDs and the standby light). If only one of these pixels is dead (i.e., always black) or stuck (fixed colour), most premium brands will replace the TV.

Liquid-crystal displays (LCDs) do not use LEDs, yet they are considered optoelectronic devices. In an LCD electric signals are used to influence the orientation of liquid crystals. By applying an electric field, the alignment of the liquid crystals is manipulated, which affects how light passes through them. The light comes from either a backlight (usually LEDs, but not necessarily) or ambient light falling onto the display. Colours are obtained with colour filters. The main advantage of LCDs over LED-based displays is their low power consumption. Even better at this are electronic- or e-paper displays as used in e-readers and electronic labels found in e.g. supermarkets. Such displays only consume energy when their contents change. They may be slow, but they can retain an image for several years without needing a power source. The image is created by tiny electrically charged balls that align themselves to an externally applied electric field. The balls retain their orientation when the electric field is removed, and the image remains visible.

If displays had a commuter class, the 16×2 alphanumeric LCD would be its Toyota Corolla: dependable, unassuming, and practical.

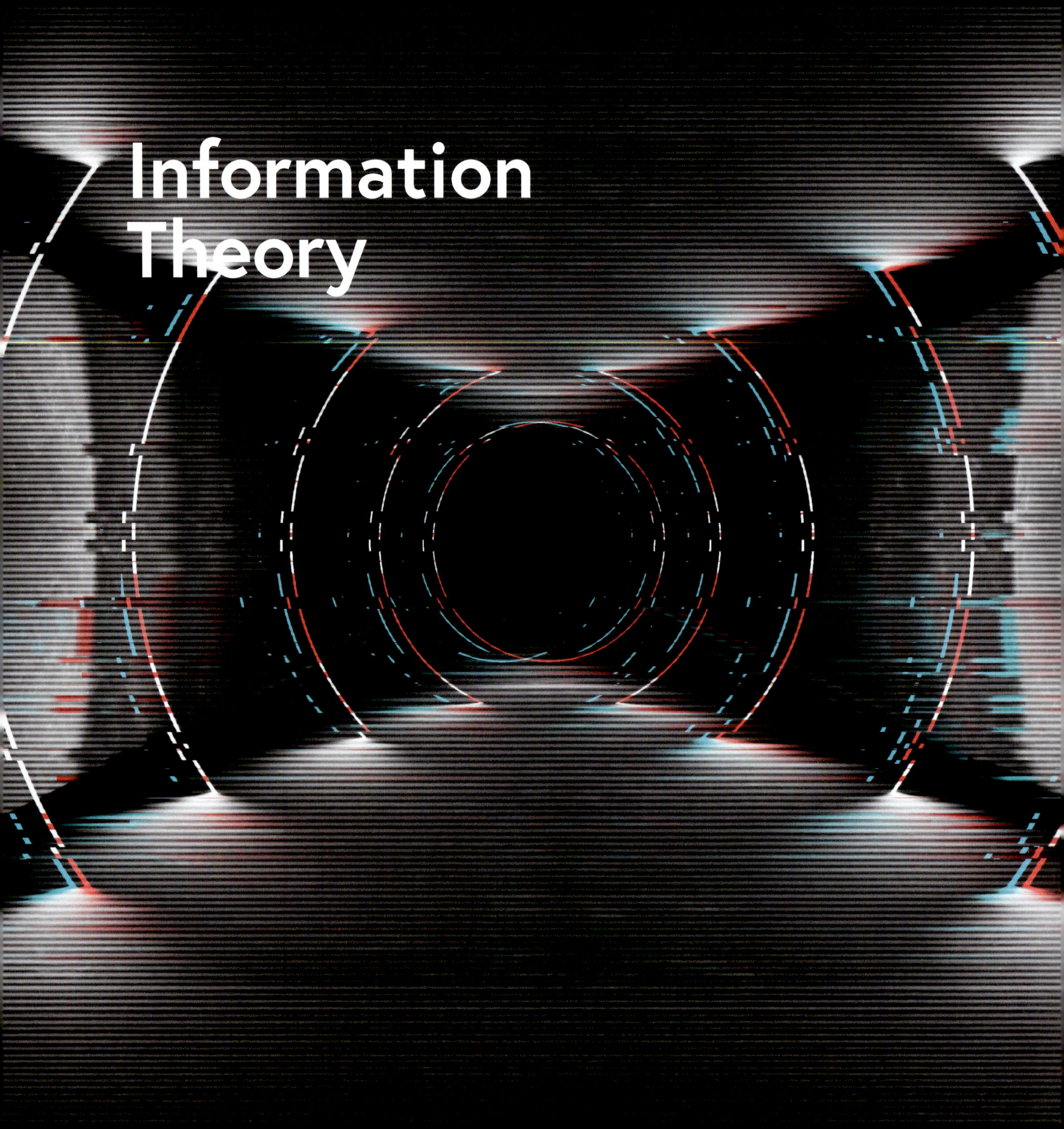
Information
Theory

American mathematician and electrical engineer Claude Elwood Shannon (1916–2001) was a brainy chap. In 1938, as a student, he showed how Boolean algebra could be used to simplify telephone exchange systems and how relays could be used to solve Boolean algebra problems. With his thesis, he marked the beginning of digital computing and digital logic circuit design. A year later, only 23 years old, Shannon was awarded the Alfred Noble Prize[22] for this work.

In 1948, Shannon published another landmark paper that laid the foundations of information theory. Today, this paper is considered one of the most influential scientific papers of all time. In it, Shannon introduced the concepts of channel capacity, the noisy-channel coding theorem (also known as Shannon's theorem or Shannon's limit) and the bit.

In 1950, Shannon built the Theseus, a self-learning robot. It could find its way out of a maze and remember the solution.
It was not only an amazing machine but also the first demonstration of machine learning (ML). A few years later, in 1956, Claude Shannon was one of the organizers and participants of the Dartmouth workshop, which is now considered the founding event of the field of artificial intelligence (AI).

Impressive as this all may be, what interests us most in this chapter is Shannon's 1948 paper. The figure shown here appears at the beginning of that paper (without the '+'-sign in the circle). Like many good ideas, it is rather simple but has far-reaching consequences.

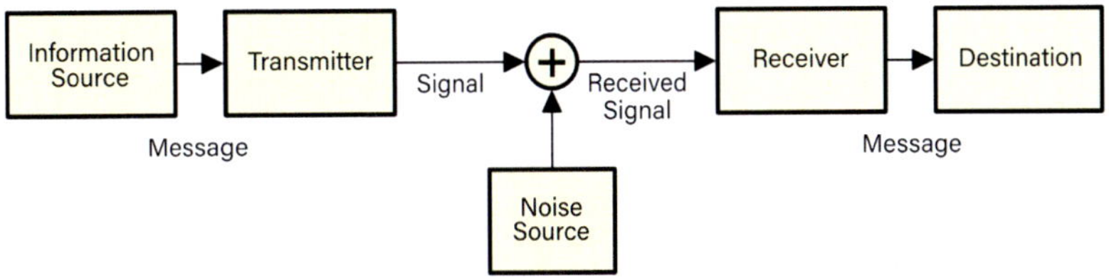

Schematic diagram of a general communication system as presented in Shannon's 1948 paper.

In a communication system, any message originating from the information source will inevitably be contaminated to some extent by noise before reaching its destination. A message can be anything from discrete characters (e.g., texting) to sets of continuous functions of multiple variables (e.g., a digital television signal). Noise in this context is a disturbing signal that alters the message in some way and thereby adds uncertainty. Shannon developed the concept of entropy to quantify the uncertainty (i.e., the noise) within a message.

22 Ah, caught you there, eh? The Alfred *Noble* prize and the Alfred *Nobel* prize are two different awards. Shannon never won the latter.

The word 'bit' is attributed to American mathematician and statistician John Wilder Tukey (1915–2000), best known for the Cooley-Tukey FFT algorithm.

He also introduced the bit[23] as the fundamental unit of information. However, a mathematician of Shannon's calibre can take such ideas much further, and he transformed them into a comprehensive mathematical framework for information.

Besides the bit and entropy, Shannon also introduced redundancy and channel capacity. Redundancy in a message is the repetition or predictability that can be removed without losing essential information. Removing redundancy from a message is at the heart of source coding, better known as data compression (used in e.g. ZIP archives and MP3 files). Channel capacity is defined as the maximum rate at which information can be transmitted over a communication channel without error, even in the presence of noise. Shannon's noisy-channel coding theorem states that it is possible to transmit information over a noisy channel with an arbitrarily low error rate, provided the transmission rate does not exceed the channel capacity.

Every system involving communication or data storage, whether it's a phone line, a computer network, or biological systems like DNA transmission can be studied within Shannon's framework. Applications from cell phones, fibre optics, and satellite communications, to hard drives, CDs, and cloud systems, from compression algorithms like ZIP, MP3, and video streaming to cryptography and secure communications, artificial intelligence and machine learning, Shannon's theories apply to them all. No wonder then that Claude Shannon is considered the "father of the Information Age".

In mathematical form, the entropy H is defined as follows:

$$H(X) := -\sum_i p(x_i) \cdot \log p(x_i)$$

Here, $p(x_i)$ is the probability of each possible outcome, a value in the range $[0,1]$[24]. Higher entropy means more uncertainty (e.g., a random string of characters), while lower entropy means less uncertainty (e.g., repeated or predictable patterns).

The mathematical definition of channel capacity is rather abstract, but it can be made more tangible by assuming that the channel is subject to additive white Gaussian

23 IEC 80000-13 defines the shannon as the information content associated with an event when the probability of the event occurring is 50%.

24 The logarithm of a value smaller than one is negative, which explains the minus sign in the equation.

noise. Now, according to Shannon, the channel capacity is given by:

$$C = B \cdot \log_2(1 + \frac{S}{N})$$

Here, C is the channel capacity (in bits per second or bps or bits/s). B is the bandwidth (in hertz), S is the signal power, and N is the noise power (S and N both in watts or volts^2, not in dB). This equation is also known as the Shannon-Hartley theorem.

Shannon did not think everything up from scratch, but he built upon Nyquist's earlier findings in the field of communication theory. Indeed, the channel capacity equation given above is an extension of Nyquist's formula for a noise- and error free channel:

$$C = 2B \cdot \log_2 M$$

Here, M is the number of discrete signal levels. If the signal is a binary signal that can take on only two values ($M = 2$), the equation simplifies to:

$$C = 2B$$

This is called the Nyquist rate. It gives the maximum number of code elements per second that can be unambiguously resolved at the receiver's side, i.e., after going through the communication channel.

The Nyquist rate relates to signalling and is not the same thing as the Nyquist frequency, which applies to sampled signals. In 1928, Nyquist showed that the required bandwidth to transmit a signal is directly proportional to its speed. According to Nyquist, signals with a frequency less than or equal to half the channel bandwidth can be reconstructed unambiguously. Claude Shannon extended this result to discrete-time (i.e., sampled) signals by formulating the sampling theorem. This theorem states that a sampled signal can be reconstructed perfectly if the sample rate is (at least) twice the highest frequency component present in the signal:

$$f_S = 2B$$

Here, f_S is te Nyquist frequency (in hertz). It is this simple equation that allows us to enjoy high-quality streaming audio and video and clearly hear (which does not mean understand) what the person on the other end of the phone line is trying to tell us.

Nyquist used the Fourier series to prove his point. The Fourier series allows expressing any periodic function as a sum of sines and cosines. Because goniometric functions are well understood, they can help make complex calculations somewhat easier. They were discovered by French mathematician and physicist Jean-Baptiste Joseph Fourier (1768–1830) when he studied the flow of heat. His discovery resulted in the Fourier transform and, later, the fast Fourier transform (FFT), considered by some as the most important numerical algorithm of our lifetime. Because the Fourier series are so important in electronics, here is the definition:

$$f(t) = a_0 + \sum_{n=1}^{\infty}\left(a_n \cos\frac{n\pi t}{T} + b_n \sin\frac{n\pi t}{T}\right)$$

Here, *f(t)* is a periodic function with period *T*, a_0 is the DC component (the average value of the function over one period *T*), a_n and b_n are the Fourier coefficients, and *n* is the harmonic number (multiples of the fundamental frequency 1/*T*).

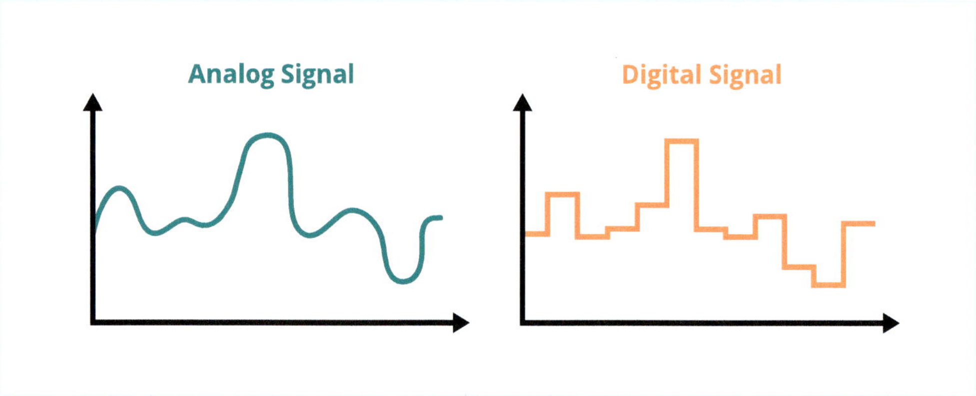

The sampling theorem has been discovered independently by several people, which is why in this book it is simply referred to as 'sampling theorem'.

The Minivac 601 — A Hands-On Introduction to Digital Logic

Claude Shannon believed that theoretical advancements should be tested and confirmed by experiment. Therefore, in 1961, he designed the Minivac 601, a hands-on device that used relays, switches and lights to teach the fundamentals of digital logic and information processing. The device was commercialised by Scientific Development Corporation as an educational toy. It was a six-bit device with a data memory of six (yes, six) bits.

The Minivac 601 was not a computer in the modern sense. It was more of a hands-on teaching tool that allowed people (especially students and engineers) to understand basic computing concepts such as binary number representation, logic circuits (AND, OR, NOT) and sequential logic, data storage, and input/output devices. It used relays to perform logic operations and to store information. It had a rotary dial and toggle switches for input, and small light bulbs for output. The Minivac 601 was programmed manually by plugging in wires and flipping switches.

Digital Logic

The digital computer is a fascinating machine. Since its official invention in 1942 it has invaded our lives. Today, almost every electronic device except for the simplest ones has some form of digital computer inside. They have become so common, that it is probably cheaper to blink an LED with a digital computer than by using a dedicated integrated circuit like the NE555 and the supporting components required for it to make it work. The success of the computer is, of course, largely due to the miniaturization of the transistor, which has allowed the integration of millions of them on a single chip and even billions in a single device. However, the computer wouldn't be nothing without a logic heart. Because that's what makes it fascinating, almost everything a computer does is based on a few basic logic functions: AND, OR and NOT.

It was English mathematician George Boole (1815–1864) who around 1850 introduced what is now called Boolean algebra. In (t)his algebra, variables can take on only two values, true or false. Multiplication is replaced by the conjunction or AND operator, while additions are done with the disjunction or OR operator. The negation or NOT operator provides negations. These three are the primary Boolean operators. Secondary operators are composed from the primary operators, a common one being the exclusive OR or XOR operator. The primary operators are easy to understand:

AND—If A is *true* AND B is *true* then the result is also *true*. The result is *false* for all other cases. The AND operator is officially written as '^' (or '•') but this is impractical in computer programming and so often the typable '&' symbol is used instead.

OR—If A is *true* OR B is *true* then the result is also *true*. Therefore, the result is *false* only for the case where both A and B are *false*. The OR operator is officially written as 'v' (or '+'). Although this symbol looks a lot like 'V', many computer programming languages prefer '|'.

NOT—If A is *true* then the result is *false*. The result is *true* when A is *false*. The unary NOT operator is officially written as '¬'. Again, in computer programming, typable symbols like '~' or '!' are preferred.

The above rules are not limited to two variables, and A and B can be compound statements.

Logical vs. Bitwise

Programming languages usually differentiate between logical and bitwise logic operations and use different operator symbols to distinguish between them, e.g., '&' for bitwise and '&&' for logical AND. Similarly, the bitwise OR can be '|' while the logical OR is represented by '||'. This is not a universal rule.

British mathematician Augustus De Morgan (1806–1871) formulated what became known as the De Morgan's Laws. They enable AND and OR operations to be expressed entirely in terms of one another through the use of negation:

NOT (A AND B) is the same as (NOT A) OR (NOT B)

NOT (A OR B) is identical to (NOT A) AND (NOT B)

These laws aren't rocket science, I could have come up with them myself, but they are practical when simplifying logic functions or circuits.

Boolean algebra did not find an application in engineering until the 1930s when Claude Shannon noticed that he could use it for analysing the switching circuits he was studying. By showing how relays can be used to solve Boolean algebra problems, he laid the foundations of digital circuit design. Later, relays were replaced by vacuum tubes, which in turn were superseded by transistors. The next step might be replacing the transistors by quantum states.

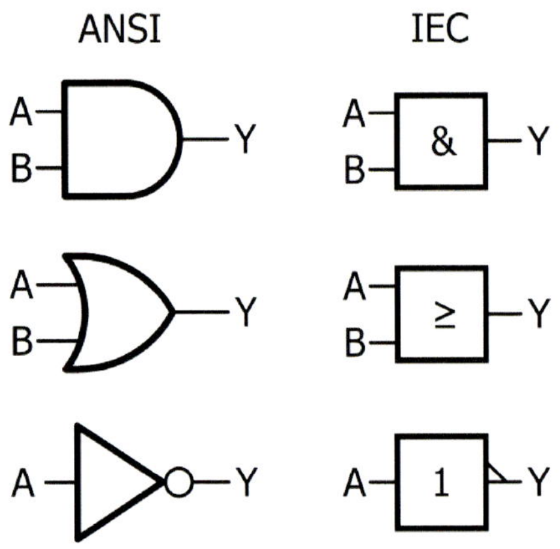

The electrical symbols for the three primary Boolean operators AND, OR (two-way) and NOT come in two flavours. Even though the IEC style allows for a more consistent and extensive representation of logic functions, many designers prefer the ANSI-style symbols.

Every Boolean function can be expressed with AND and NOT operators only. The same is true for OR and NOT operators. AND and NOT combine into the NAND operator. Similarly, OR and NOT combine into the NOR operator. Such operators are what is called functionally complete, and they are therefore also referred to as universal operators.

A question that may have arisen after reading the above might be how to use the three primary Boolean operators to build a computer? Well, as demonstrated in 1937 by American mathematician George Robert Stibitz (1904–1995)[25], one of the fathers of the digital computer, it is possible to add two binary values with nothing more than XOR operators. An XOR operator in turn can be made from NAND or NOR operators. Multiplications and divisions can be calculated by repeated additions and negations (subtraction). In other words, Boolean logic can be used for performing calculations.

25 George Stibitz is also credited with coining the term digital electronics. Until then, it was called pulse electronics.

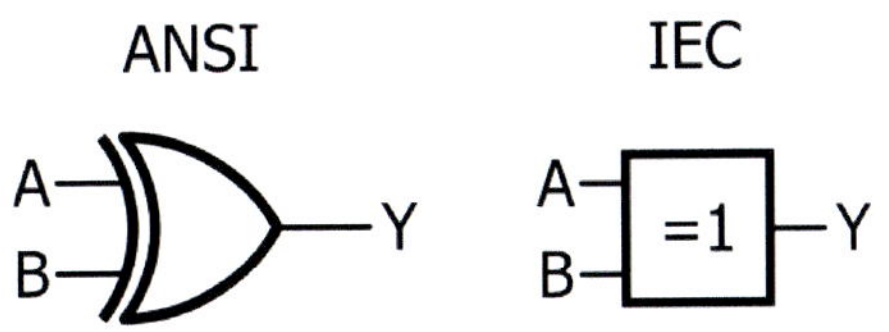

A	B	Y=A⊕B
0	0	0
0	1	1
1	0	1
1	1	0

The Exclusive OR or XOR

The exclusive OR or XOR operator (pronounced EXOR), is a secondary Boolean operator that can be made from the primary operators. However, it is such a commonly used function that it has its own mathematical symbol, ⊕, and electrical symbol. The output of an XOR gate with two inputs is *true* only when both inputs are unequal, it is *false* otherwise. This leads to an interesting problem when extending the gate to three inputs or more. Three (or more) variables that each can take on only one of two values cannot all be unequal to each other. As a result, there are two opinions of how an XOR gate with more than two inputs should behave. One school of thought claims that the result of such a gate should be true only if exactly one of the inputs is true. The more commonly accepted behaviour is that the result should be *true* only when an odd number of its inputs is *true*. This is the way the 3-input XOR gate 74AUP1G386 from Nexperia works.

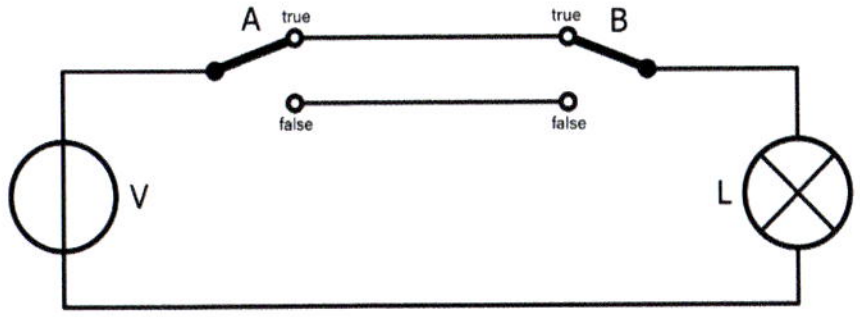

Commonly used in staircases, hallways, or large rooms where control from multiple points is convenient, this circuit allows the light L to be controlled from two different locations *A* and *B*. It is the electrical equivalent of the Boolean XNOR function, the negated version of the XOR: *L* = NOT (A XOR B). If you make the two wires between the switches *A* and cross, the circuit will show XOR behaviour.

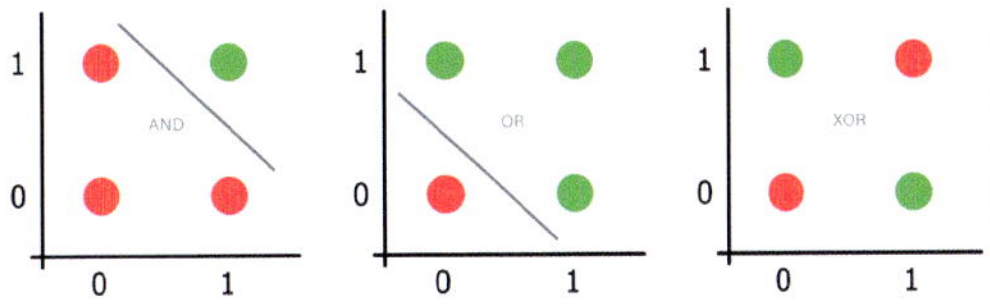

The outputs of the AND, OR and XOR functions plotted as two-dimensional graphs. The horizontal axis is input *A*, the vertical axis is input *B*. A green dot represents *true* (or 1), a red dot means *false* (or 0). In the XOR plot, it is not possible to separate the red dots from the green by drawing a single straight line. Therefore, the XOR function is not linearly separable.

The XOR function also plays an important role in artificial intelligence and machine learning. This is due to the fact that when plotted as a two-dimensional graph, the XOR output cannot be separated into *true/false* regions using a single straight line, making it a so-called non-linearly separable problem. Simple learning systems can only solve linearly separable problems like AND and OR. Therefore, the XOR function is used as a benchmark for machine learning algorithms and neural networks. If an algorithm cannot solve non-linearly separable problems, it is not really suitable for complex AI applications.

As a side note, after inventing the binary adder, Stibitz filed a patent for a binary counter (1941), and he built the Complex Computer (1944). Entirely based on relay logic, this computer performed calculations with complex numbers. It was quickly extended with multiple input terminals, introducing time-sharing. George Stibitz also used his Complex Computer to demonstrate remote computing.

Combining logic functions to create more complex functions and perform Boolean algebra is called combinatorial logic. Albeit one of the cornerstones of a modern computer, it is not sufficient for building a computer. A computer also needs to store data and memorize temporary values in between calculation steps. This is where sequential logic enters the arena. In sequential logic, the output of a function depends not only on the current values of its inputs but also on previous states. This implies that some form of memory is required to retain intermediate states.

The basic building block of computer memory is the flip-flop, also known as the bistable multivibrator. It was invented in 1918 by British physicists William Henry Eccles (1875–1966) and Frank Wilfred Jordan (1881–1941). They didn't think of it as a binary memory, but as a kind of electronic relay. Initially it was called the Eccles-Jordan trigger circuit but it entered history as the bistable multivibrator (officially) or flip-flop (common language). Years later, it was recognized that such a circuit could be used as a single-bit binary memory in computers. Today, we integrate millions of them on a single memory chip.

Eccles and Jordan built their flip-flop out of vacuum tubes, but today transistors are used, of course. NAND gates, themselves consisting of transistors, can be used too. Only two gates are required to build a simple flip-flop. NAND flash memory for instance, used in solid-state disk drives like SSDs and pen drives, is based on this technique. NOR gate-based flash memory exists as well.

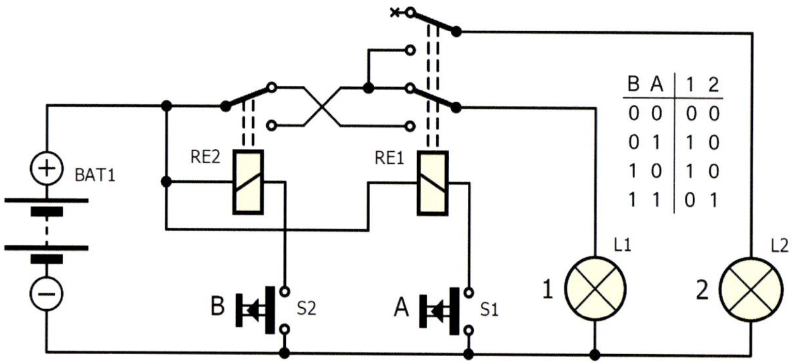

B	A	1	2
0	0	0	0
0	1	1	0
1	0	1	0
1	1	0	1

The 2-bit binary adder circuit designed by George Stibitz used relay logic. Lamp 2 is the carry-over bit or carry. Such a circuit is called a half-adder because it does not take into account any carry over from a previous addition. Lamp 1 is connected to the output of the XOR gate created with the switch of RE2 and the lower switch of RE1. Lamp 2 is connected to the output of the AND gate formed by the switch of RE2 and the upper switch of RE1. The truth table shows the output and input signals. Cascading XOR operators allows adding values of any word size (i.e., bit length).

Please note that as I couldn't find a schematic of Stibitz's device, I reconstructed it from photographs of reproductions of his prototype found online. Therefore, there may be mistakes (even though it works as intended).

The Intriguing Case of the Multivibrator

Ever since I started out in electronics I have been amazed by the word 'multivibrator', though I never went through the trouble of trying to find out how it got its name. I just accepted it as a fact. For this book I decided to repair this lack of thoroughness and go to the bottom of the multivibrator case.

The multivibrator was invented during World War I by French physicists Henri Abraham (1868–1943) and Eugène Bloch (1878–1944). In those days, oscillators generated sinewaves, and they vibrated at a single frequency. The new device, on the other hand, produced a square wave. As Fourier theory tells us, such a waveform is rich in harmonics where each harmonic can be considered a sinewave vibrating at the harmonic's frequency. Therefore, the new oscillator vibrated at multiple frequencies and was dubbed a multivibrator[26]. Unfortunately, this term stuck, making life difficult for future engineers by imposing weird terminology.

The multivibrator is referred to as 'astable' because it continuously toggles between two states. Eccles and Jordan modified the circuit originally developed by Abraham and Bloch, making it stable in either of its two states. This modified version, known as the bistable multivibrator, became the modern flip-flop. From these circuits, a new variation eventually emerged: a device stable in only one state. This device, known as the monostable multivibrator, can be temporarily switched to its unstable state, but it automatically returns to its stable state after a certain period. This is the monostable multivibrator or monoflop a.k.a. one-shot. This type of circuit is commonly used in timing and pulse-shaping applications. All three types of multivibrators can be realized with NAND or NOR gates.

The multivibrator, in its various forms, has become a widely used device in electronics. All types can be found in applications such as clocks, computers, radios, automobiles, electronic musical instruments, kitchen appliances, and many other devices.

26 Following this reasoning, a sinewave oscillator should be called a monovibrator.

Digital logic ICs with their power supply decoupling capacitors.

With logic gates and flip-flops, a wide range of interesting functions can be created, including counters, multiplexers, latches, encoders, decoders, comparators, shift registers, and more. Many of these common functions have been integrated into building-block integrated circuits, which can be used to construct more complex devices, such as computers. In the 1970s, digital systems flourished. Circuit boards containing dozens of digital logic chips were common and systems often consisted of several of such boards. It was during this period that the now ubiquitous decoupling capacitor made its appearance. It was found that to ensure proper operation of each logic circuit, it was necessary to provide power supply filtering for every IC on the board. This quickly became common practice. Today, electronics engineers routinely place decoupling capacitors close to the power supply pins of every IC they use in their designs, even if it isn't really necessary. It is simply considered good design practice.

Since digital logic is so versatile, limiting its applications to preconfigured ICs and microprocessors would be wasteful—or so believed American engineer Ross H. Freeman (1948–1989) in the early 1980s. In 1984, he invented the field-programmable gate array (FPGA). Commonly referred to as FPGAs, these are programmable devices containing numerous logic gates and flip-flops that can be reconfigured after manufacturing. They allow engineers to adapt designs to changing specifications or add new functions, offering flexibility that standard ICs and application-specific ICs (ASICs) cannot provide. Moreover, FPGAs offer a faster and more cost-effective way to prototype, test, and implement designs. While FPGAs have since evolved into a multi-billion-euro market, Ross Freeman did not live to witness their success. He passed away at the age of 41, less than a month after his first patent application was approved.

As of 2025, FPGAs can integrate up to twenty million logic units, each consisting of multiple logic elements[27]. In addition to a massive logic canvas, FPGAs can include features like microcontroller cores, high-speed communication interfaces, and significant amounts of memory. These powerful devices are well-suited for applications such as high-performance parallel computing and various hardware accelerators. They also provide testbeds for new microprocessor designs.

Let's terminate this chapter on logic with a word on types of logic. We already came across relay logic and digital logic, but there are more ways to implement logic functions. Closely related to relay logic is, of course, switch logic as shown in the section on the XOR. Transistors are a kind of electronic relays and so it should not come as a surprise to learn that transistor logic exist also. More surprising may be to learn that diodes too can be used to implement AND and OR gates. Diode logic has the particularity that it can easily mix signals of different levels. As an example, the diode OR circuit is often used in power supply selection systems to choose the highest available supply voltage or to determine if an external power adapter is connected, allowing the battery to be disconnected.

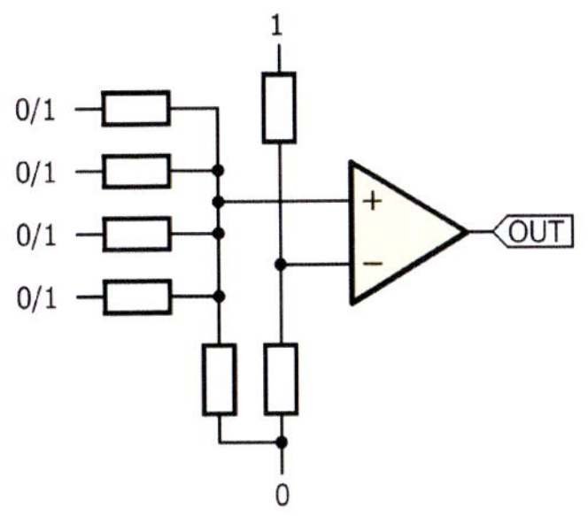

In the above circuit,, when all the resistors have the same value, the output of the comparator will be high when a majority of the inputs are high, i.e., three or four. It works as follows. The voltage on the inverting input of the comparator is 0.5 V, while the voltage on the non-inverting input depends on the states of the inputs. When all inputs are '0', the voltage on the non-inverting input is 0 V. When all inputs are '1', it is 0.8 V. Each '1' adds 0.2 V (one-fifth) to the voltage on the inverting input. Therefore, when two inputs are '1', the voltage is 0.4 V. This is lower than the voltage on the inverting input, so the output of the comparator is '0'. When three inputs are '1', the voltage reaches 0.6 V, which is higher than the voltage on the inverting input, causing the output to become '1'. Thus, the circuit functions as a majority gate. When the output is inverted, it becomes a minority gate.

27 According to AMD, their Versal VP1902 FPGA launched in 2023 supports designs up to 60 billion gates. A gate translates to approx. four transistors...

The Microcontroller

The miniaturization of the transistor allowed for building complex logic circuits on a single chip. The first commercially produced central processing unit (CPU) or microprocessor (μP)—the brain of a computer, the heart is its clock oscillator—was released by Intel in 1971. It was a 4-bit device named 4004. It was revolutionary in at least two ways.

First of all, it was a universal device that could execute programs stored in external memory instead of being dedicated to a single function (a desk calculator, as a matter of fact). The second innovative feature of the 4004 was the use of the silicon-gate technology (SGT). Developed by Italian physicist and engineer Federico Faggin (1941–) at Fairchild Semiconductor, silicon-gate technology allowed him to create the high-density silicon design that fitted all the required logic functions on a single chip. The 4004 integrated the then staggering number of 2,300 transistors.

Also in 1971, and even a few months before Intel introduced the 4004, Texas Instruments launched the TMS-0100, a calculator on a chip. This device integrated not only a CPU, but also memory (RAM, random-access memory) and read-only program memory (ROM, read-only memory). Even though its program was baked into the chip, making it effectively a special purpose device, its design allowed universal usage. Today, the TMS-0100 is considered the world's first microcontroller (μC or MCU, microcontroller unit). In 1996, following the customary patent disputes, American engineer Gary W. Boone (1945–2013) was officially recognized inventor of the microcontroller, for which he had filed a patent in 1971.

As will be clear from the above, there was a distinction between microprocessors and microcontrollers right from the beginning. Initially, the difference between the two was in the application domains: special purpose applications for the microcontroller versus universal applicability of the microprocessor. Both continued to evolve into ever more powerful devices. Microprocessor development concentrated on increasing computing power whereas microcontrollers gained more and more peripherals and got larger memories. Computing power too has increased. As of 2025, microcontrollers can be had from 4-bit, 6-pin devices up to 64-bit devices with hundreds of pins. They are deployed in almost everything, from smartphones and coffee machines to drones and missiles (to name a few popular applications). Microprocessors on the other hand still concentrate on providing computing

Federico Faggin (co)designed several well-known processors like the 8080, Z8 and Z80 (shown here). He was also closely involved in the development of the touchpad, commonly found in modern laptop compu

power. They are found mainly in computers, from laptops to supercomputers.

With the ongoing miniaturization, computer systems, both microprocessor and microcontroller-based, evolved from stand-alone systems into building blocks that can be integrated into a larger system. Eventually, this led to embedded systems or electronics becoming a separate discipline of electronics at the beginning of the 21st century. Compared to the classically trained electronics engineer, your typical embedded systems engineer has both computer programming and electronic design skills, and (s)he focuses on interfacing the virtual digital world to the analogue real world.

As microcontrollers gained in performance, the role of software became more prominent. Graphical user interfaces and powerful multitasking operating systems transformed the embedded system into a computer running one or more specialised tasks. A modern TV set or smartphone has become a computer that executes an operating system and runs applications ('apps'). They allow the connection of a user input device (i.e., a keyboard) or have one built in, and they can connect to the internet. From radios with a computer inside, they have become computers with a radio inside.

A similar evolution has occurred in the opposite direction, towards the analogue world. Microcontrollers have found their way into nearly every conceivable application, where they are used to measure physical quantities and control processes. To give you an idea, tens of billions of microcontrollers are sold each year[28]. They are simply everywhere. (Where else can they be?) These microcontrollers are also becoming increasingly powerful and capable of connecting to the internet. Communicating over the Internet of Things (IoT), they exchange information that enables them to

28 The estimate for 2025 was over 40 billion physical devices. This number is much higher when soft cores that go into programmable logic are counted too.

control processes spanning large areas, utilise artificial intelligence, or both—or even perform something as simple as blinking an LED.

Pure analogue electronic circuits (as opposed to embedded systems) are still essential for interfacing with physical quantities such as temperature, humidity, strain, light, and more. These circuits amplify the signals from sensors before converting them into digital data that can be processed by a microcontroller. In the past, additional analogue circuitry was often used in the signal chain to perform mathematical operations such as integration or filtering, which were necessary for process control. However, these operations can now be implemented in software executed by microcontrollers. This not only simplifies analogue circuit design but also increases system flexibility.

A striking example is radio (in electronics, radio is the proverbial butler of a whodunnit novel), where software has eliminated approximately 90% of the analogue domain. In software-defined radio (SDR), the software handles modulation and demodulation, up- and down-conversion, and other purely mathematical operations that were once implemented with transistors and passive components. Only the input and output stages, such as amplifiers and antennas, have remained analogue.

The advantage of SDR is evident: any modulation scheme can be implemented on the same platform, provided that sufficient computing power and analogue input and output bandwidths are available. A single system can not only perform tasks such as FM radio reception (soon to be obsolete?), decoding TV signals, or connecting to a Wi-Fi or mobile phone network, but also implement algorithms that have yet to be invented.

The SDR paradigm is extending into other domains, such as measurement instruments and music, where it is used to emulate analogue vintage equipment as well as to implement new sound generation and processing algorithms. In other words, special-purpose applications that used to integrate a computer are becoming computers with special-purpose interfaces.

With tens of billions of devices sold each year, in such an important market with so many applications, it isn't surprising that about every semiconductor manufacturer tries to grab a share of it. This is where microcontroller design companies enter the arena. Such companies create microcontrollers on paper and sell these to others

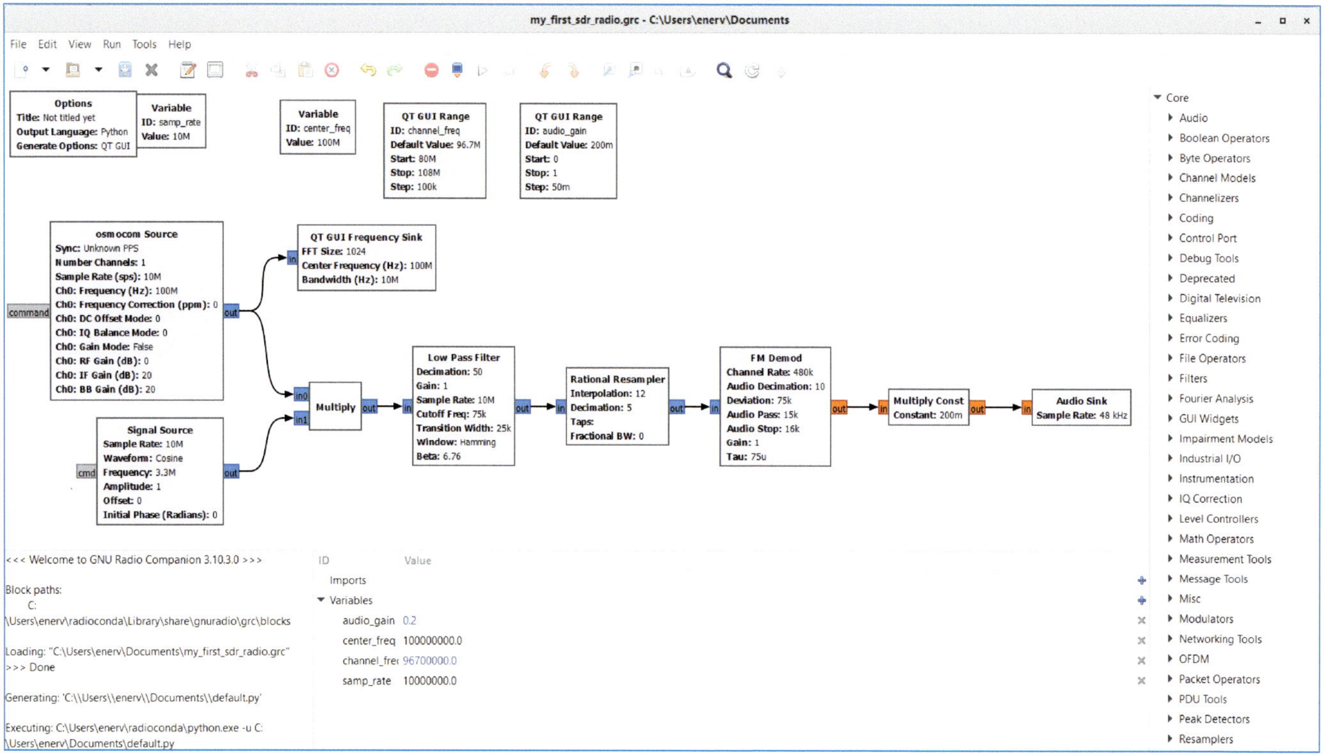

A software-based implementation of an FM radio receiver with tune and volume controls created in the open source SDR toolbox GNU Radio.

who put them into silicon and manufacture physical devices. This way, you too can become a player in the microcontroller market (if you have the money to back up your ambitions).

The first microcontroller available from multiple sources was the now classic MCS-51 released by Intel in 1980. Its instruction set architecture (ISA), designed by American engineer John Harrison Wharton (1954–2018), entered the hall of fame as the 8051. The key innovation of the design was its instruction set, which decoupled the computer core from its peripherals. Before this, every new microcontroller or variant introduced a slightly different instruction set to support its specific features. This required users to adapt their software for each new device—a process that was time-consuming, error-prone, and costly.

In the 8051, John Wharton addressed this issue by combining an immutable compute core with a universal interface for the built-in peripherals. This approach allowed new peripherals to be added later without requiring any knowledge of their specifics and without necessitating changes to the device's instruction set. To the user, a new peripheral—regardless of its type—was simply a set of additional registers to read data from and to write to. Existing software remained functional, and only the new functions needed to be implemented and tested, saving significant time and money. Similarly, licensees of the 8051 could add peripherals without having to modify the core itself.

A second reason for the success of the 8051 was Intel's ability, as a large company, to produce tools, documentation and training materials for its products. Licensees could simply direct their clients to Intel's resources instead of creating their own. This allowed them to focus on offering slightly cheaper or better-adapted versions of the 8051 to their clients. The only documentation they needed to provide was how to use the features they had added. Therefore, even a small company could quickly become a player in the juicy microcontroller market. The 8051 and its siblings are used in toys and home appliances, just to name a few high-volume applications.

Twenty years after its introduction—the typical lifespan of a microcontroller—the 8051 defied obsolescence, with around a thousand variants actively competing for users and applications. Even though Intel discontinued production of the 8051 in 2006, many companies keep producing it. Additionally, open-source soft-core designs have become available, effectively allowing anyone to incorporate an 8051 into a product virtually for free. Nearly fifty years after its debut, the 8051 remains very much alive and thriving, with no signs of fading away anytime soon.

Block diagram of the 8051. Even though there are many blocks and arrows, it is a relatively simple microcontroller design.

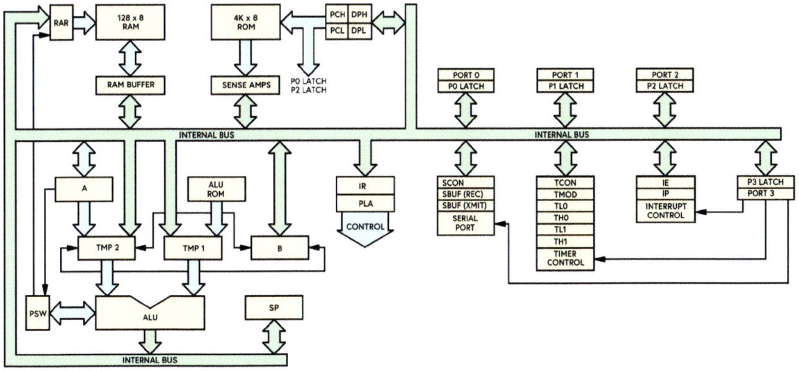

No matter how successful the 8051 was—and still is—it was designed in 1978, and technology has evolved since then. By the 1980s, high-level programming languages like C and Pascal were gaining popularity and began to be used for developing microcontroller applications. C, created in the 1970s by American computer scientist Dennis MacAlistair Ritchie (1941–2011)[29],

Nearly fifty years after its debut, the 8051 remains very much alive and thriving.

29 Denis Ritchie is also one of the creators of the UNIX operating system.

Okay, but What the Heck is a Microcontroller?

A microcontroller is a self-contained computer in a single integrated circuit (IC). It combines one or more processing cores with program and data memory and a set of peripherals in a compact package. This contrasts with general-purpose computers, which typically require multiple chips. Microcontroller peripherals are tailored to specific applications and can include communication interfaces, timers, analogue-to-digital converters, and motor controllers. The peripherals of a general-purpose computer are usually limited to high-speed ports for data communication and video.

Microcontrollers come in a wide range of packages, from as few as six pins to several hundred. Word widths range from 4 to 64 bits or more, and memory sizes span from a few kilobytes to several megabytes. Most use some form of the Harvard architecture, which separates data and program memory.

One of the features behind the microcontroller's success is electrically erasable program memory (Flash or EEPROM), which allows for in-circuit programming and remote updates, including over-the-air (OTA) firmware changes. This is vital for maintaining devices that are difficult to access, such as (deeply) embedded IoT nodes.

A microcontroller contains most of what is needed to run an application, including the software. When a microcontroller integrates about everything required for a system, it is called a System-on-Chip (SoC). In some cases, microcontrollers include other chips (like memory) within the same package, becoming System-in-Package (SiP) devices.

has since become one of the most widely used programming languages. However, the architecture of the 8051 does not align well with the standard C language. To program the 8051 in C, language extensions are required, resulting in source code that is difficult to port to other platforms—and vice versa.

Other microcontroller architectures, better suited to embedded applications and system-level programming languages like C, began to emerge. One of the companies designing such cores was the UK-based company ARM. By 2010, one of their objectives was to achieve for their 32-bit products what Intel had accomplished with the 8051[30]. Fifteen years later, it is clear that they have succeeded in this mission. With their Cortex core designs licensed to nearly every semiconductor manufacturer on the planet, ARM's estimated share of the global microcontroller market is now approaching 50%.

The ARM core instruction set architectures (ISAs) are proprietary, like the ISAs of all other commercial microcontrollers, and therefore cannot be used for free. The open-source community has developed several open-source ISAs, but none has yet achieved the success of Intel or ARM. The primary reason for this has been the lack of serious support, which made their use in commercial applications risky.

However, in 2015, the RISC-V Foundation was established to address this issue. Renamed RISC-V International in 2020, this non-profit organisation promotes the RISC-V ISA—the fifth-generation specification for a reduced instruction set computer (RISC[31], as opposed to a complex instruction set computer or CISC). RISC-V is now actively supported by many major microcontroller designers and manufacturers, and RISC-V-based cores are becoming increasingly widespread. It may, however, take a few more years for RISC-V to achieve world dominance—if it ever does. Only time will tell.

30 As explained during a roundtable meeting with ARM, Keil, and NXP at the ARM European Technical Conference held in Paris, France, on October 21, 2010, where I happened to be present. (Excellent food, by the way.)

31 Some consider the famous 6502 microprocessor from 1975 a reduced instruction set processor, even though the word RISC didn't exist yet.

Digital Signals

In an analogue electronic system, signals consist of currents and voltages that fluctuate between a minimum and a maximum value. A value of a physical quantity or a parameter is represented by a voltage or current somewhere in between these two limits. A good understanding of how to process and handle such signals is required when designing a device or system. For example, when input and output impedances don't match, signals may be affected and some of the information conveyed by them may be lost or altered. As a result, the system may not work as intended.

In digital electronics signals are also transported by voltages and currents, except that they can usually take on only two values: low and high, or '0' and '1'. This is the case for binary systems. Systems using three or four or more values exist too, but in this chapter, we will concentrate on binary signals. When a signal can take on only two values, a method must be devised to transmit information that requires more than two values. Indeed, how do you inform a microcontroller about the position of a switch that has three positions or more using only two values? (Spoiler alert: introduce another dimension like time or space.)

This is where communication protocols come in. To put it formally in a lengthy way, a communication protocol is a set of rules and standards that define how devices exchange data over a network or communication channel. It ensures that devices can send, receive, and interpret messages correctly, regardless of differences in hardware, software, or location. A well-defined communication protocol ensures that different devices and systems can interoperate reliably in a structured way. In short, two systems must speak the same language to understand each other.

Humans have developed many languages. At the time of writing, there are approximately 7,100 living languages worldwide. Humans being humans, they created about the same number of communication protocols (and programming languages, and connectors, end what not) for their digital systems. Much like the Babel of human languages, the tech industry is filled with a chaotic mix of communication protocols, each struggling to be understood. Like human languages, some protocols are more widely spoken than others, and dialects exist too.

In 1840, American inventor and painter Samuel Finley Breese Morse (1791–1872) obtained a patent for a method of transmitting an alphabet of characters over a single wire using a kind of pulse coding. Morse had solved the problem of encoding a set of characters using only two values by adding the dimension of time. By dividing the characters into different series of pulses and pauses, a.k.a. marks and spaces, or dots and dashes, a message could be transmitted over a single telegraph wire.

Morse code was one of the first serial communication protocols and many have followed since. In a serial protocol, each character of a set or alphabet is encoded as a unique series of ones and zeroes that are transmitted one after the other over a single wire. Synchronisation of the transmitter and the receiver is critical in this kind of asynchronous communication where the sender and receiver are 100% independent of each other.

Even though Morse code was practical and widely used, it also had a major inconvenience: it was complicated to decode mechanically, i.e., with a machine. Therefore, alternatives were sought. In the 1870s, French telegraph engineer Jean-Maurice-Émile Baudot (1845–1903) developed a telegraph alphabet where each character consisted of five bits. By using a special shift mode, he could encode the characters 'A' to 'Z', the numbers '0' to '9' and a collection of special characters[32].

In Baudot's coding system, the transmission of a character started by sending a start bit (always low), followed by the five character (data) bits and terminated with one or more stop bits (always high). As all the characters had the same length, automatic decoding was much easier than for Morse code. The data rate was defined before the transmission of a message and remained constant throughout.

Modern asynchronous serial communication protocols still use start/stop bits and fixed symbol rates for synchronization,

32 As of 2025, French (azerty) computer keyboards still use shift mode for numbers.

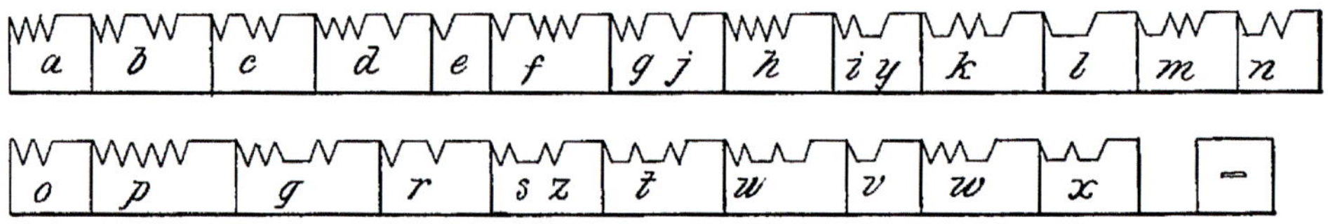

Morse code, as originally conceived by Samuel Morse in 1840, used dashes of different lengths as well as gaps, as seen in the representations of letters such as 'F' and 'T'. In this system, SOS would have been encoded as dit-dah-dit—dit-dit—dit-dah-dit rather than the familiar dit-dit-dit—dah-dah-dah—dit-dit-dit. The reason for this difference is that Morse code was standardized in 1865, leading to the development of International Morse Code.

just as Baudot did. The symbol rate is expressed in baud (Bd), after Baudot. Baudot code however did not survive and has been replaced by more elaborate schemes capable of encoding more characters like ASCII and Unicode. Published in 1963, ASCII uses seven bits per character and is probably the best-known character encoding scheme. Unicode, a more ambitious encoding standard, was designed to support the use of text in all of the world's writing systems that can be digitized. To do so, it needs more than one byte per character.

In 1960, the Electronic Industries Association (EIA) published a Recommended Standard that became famous in electronics: RS-232. The standard defined electrical signalling, pin configurations, voltage levels, and handshake mechanisms for a serial communication link but not how data is to be transmitted. It went though several iterations, the latest being TIA-232-F published in 2012 (in 1988, the EIA was renamed to TIA). With the advent of the computer, RS-232 became the standard of choice for interconnecting systems. The data format most commonly adopted was similar to the one used in Baudot code, but with eight data bits instead of five. Sometimes, a so-called parity bit is added to allow detecting transmission errors.

A 25-Way Connector for a Single-Wire Link

The RS-232 standard specifies nearly every aspect of a serial communication link except for the actual method of data transmission. It defines the connectors to use, wiring configurations, maximum cable length, and the number of required wires, among other things. To enhance robustness and flexibility, additional signals were introduced alongside the single data line in Baudot's system. In reality, RS-232 includes two data lines—RX (receive) and TX (transmit)—which enable full-duplex, two-way communication. These are supplemented by up to eighteen signals, as well as ground, shield, and power connections. Some signals allow devices to detect whether another device is connected, and if so, to wake it up when a message needs to be transmitted. A 25-pin (DB-25) connector is used to accommodate all these signals. In fact, a fully populated 25-pin RS-232 link features two independent full-duplex serial communication channels, which can operate in either synchronous or asynchronous mode and is hardly ever used.

The E-type subminiature (E-sub) connector shown at the bottom left is widely used in the industry for serial communications between computers and machines. It is usually referred to as DB-9, but its correct appellation is DE-9. The 25-way version is a B-sub type and therefore calling it DB-25 is correct. In the lower-right corner sits a 15-way DA-15 connector.

Initially, dedicated integrated circuits were used to serialize and deserialize RS-232-style communications (the so-called Universal Asynchronous Receiver/Transmitter or UART), but with ongoing miniaturization, they ended up inside the microcontroller where they became known as serial ports. Today, virtually every microcontroller features at least one serial port.

RS-232 allowed two physically separated systems like a computer and a modem to talk to each other. However, inside systems a need for digital communication signals arose too. In the 1980s, in a desire to simplify wiring in colour TVs, Philips developed a communication protocol for integrated circuits. It was called Inter-Integrated Circuit or I^2C (a.k.a. I2C and IIC) and used two wires. One wire was for the synchronization signal (clock) and the other for the data. I^2C not just defines how to send the data but also adds a high-level master-slave protocol with error detection and other features. When the patents expired in 2001, it found widespread adoption. Like the serial port, today almost every microcontroller is equipped with an I^2C bus[33]. In 2014, work started to improve I^2C, which resulted in I3C. At the time of writing, I3C is starting to find its way into microcontrollers and applications.

A main limitation of both the serial port and the I^2C bus is the maximum data speed they allow for. This makes them unsuitable for applications that require fast transfers of large chunks of data like memories and displays. The Serial Peripheral Interface or SPI, developed in the 1980s by Motorola, addressed this issue. Original SPI comprises four wires, it is a bus, and allows sending and receiving data simultaneously (i.e., it is full-duplex, as opposed to half-duplex for I^2C). Initially, speeds were in the range of a few hundred kilobits per second but today, high-speed SPI allows speeds of 100 Mbps and more. Data transfer bandwidth can be increased by adding more data lanes. Quad-SPI (QSPI) uses four data lanes and Octo-SPI eight. Again, virtually every modern microcontroller features at least one form of SPI port

When more than one data lane is used in a communication protocol, serial communication transitions into parallel communication. This method allows multiple data bits to be transmitted simultaneously by adding a separate conductor (lane) for each additional bit. Computers use parallel buses to connect the microprocessor (CPU)

33 I^2C is called a bus, because it uses more than one wire.

to other integrated circuits (ICs), such as memory chips and communication devices. Older PCs commonly featured a parallel port for connecting to printers. Internally, microcontrollers use buses to link the CPU with memory and peripherals integrated into the device. Similar to the serial link, the parallel bus can be synchronous (using a clock signal) or asynchronous.

To solve the problem of encoding a parameter that can take on more than two values by using only high and low or '1' and '0', parallel communication uses the dimension of space to add more wires. Remember, serial communication uses time for this.

In a sense, radio was at the origin of serial communication like it was at the origin of almost anything else in electronics. Digital electronics now returns the favour as parallel communication techniques have spread back into RF. Traditional radio communication systems use a single antenna at the transmitter and receiver. This is called Single-Input Single-Output or SISO. The increasing computing power of digital technology together with the demand for faster wireless data communication resulted in the exploration and development of Multiple-Input Multiple-Output (MIMO) radio systems that employ multiple antennas on both ends. MIMO, sometimes using

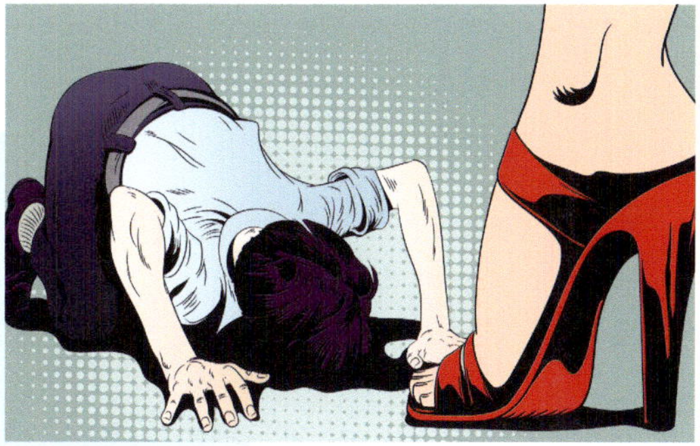

The Master/ Slave Controversy

In electronics and computing the "master/slave" terminology has historically been used to describe hierarchical relationships where one device (master) controls another (slave). The slave follows the master's commands or operates in response to its signals. Communication protocols like I^2C and SPI use master/slave roles to define primary and secondary devices. While the master/slave terms never referred to human slavery in technical contexts, some people associate it with slavery and consider it unnecessary when more neutral alternatives could be used. In 2003, Los Angeles County asked vendors to stop using master/slave in computer systems. Since then, many neutral replacements for master/slave have been proposed, like primary/secondary (as used for transformers), leader/follower, main/replica or requester/responder. In 2017, the I3C bus standard officially replaced master/slave with controller/target, a change that also made its way into version 7 of the I^2C standard.

USB-PD concerns power supplies, not communication standards and therefore doesn't fit in this chapter. As of 2025, it's in its fourth revision.

hundreds of antennas (massive MIMO), dramatically increases network capacity, efficiency, and coverage.

As requirements constantly evolve, communication standards are adapted and created to satisfy the new demands. Until now, RS-232 gave birth to RS-422(A) , RS-423, RS-449, RS-485(A) , RS-530, and RS-562. The Universal Serial Bus, better known as USB, appeared in 1996 mainly to get rid of all the different cables connecting a computer to its peripherals. Its universality turned out to be relative; with USB4 version 2 and five different connectors, it entered its tenth revision in 2022. USB competed with FireWire and won, but Thunderbolt version 5 might very well replace USB (so much for universality). Ethernet has evolved from 10 Mbps (1980s) to 1.6 Tbps, supporting various copper (twisted-pair) and fiber-optic implementations. HDMI has continuously improved from 1080p (HDMI 1.0) to 8K/10K (HDMI 2.1a), adding HDR, high refresh rates, and gaming features. The latest HDMI 2.1a standard supports 48 Gbps bandwidth for high-performance displays. HDMI however is being surpassed by DisplayPort (now in revision 2.1)

Besides digital communication standards that compete for bandwidth, there are also many others that serve different purposes. I²S, for instance, used for transmitting digital audio between ICs inside devices, or S/PDIF which is for transporting digital audio between devices. Or the Controller Area Network (CAN), initially developed by Bosch to simplify wiring in cars. Today, five variants exist. FlexRay was developed to overcome the shortcomings of CAN, whereas LIN (Local Interconnect Network) was created to simplify CAN. Similarly, the System Management Bus (SMBus or SMB) is a specialized, scaled-down version of I²C. Because it was too specialized for certain applications, the Power Management Bus (PMBus) was created.

SPI exists in several flavours and many SPI-like custom protocols have seen the light of day. Currently popular are those used to control addressable RGB LEDs. This brings us to DMX512, a digital communication protocol used for controlling stage lighting, effects, and other theatrical equip-

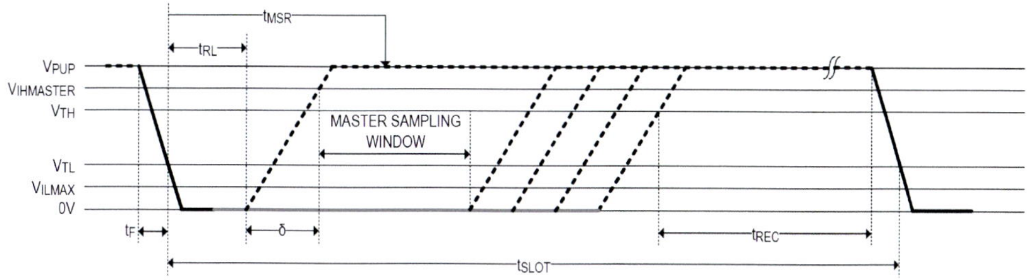

The timing diagram for a read-data time slot on the 1-wire bus, yet another serial protocol. It specifies voltage level setup- and hold times that must be respected for a transaction to succeed. This diagram was taken from the datasheet of the Analog Devices DS28E18, a 1-wire to I^2C/SPI bridge.

ment. Musical instruments, on the other hand, communicate over MIDI. Introduced in 1982, revision 2 was unveiled almost 40 years later, in 2020. X10 is a standard for home automation, but wireless protocols like Wi-Fi, Bluetooth and Zigbee have taken over.

This list could go on forever, but I'll stop here—I think you get the idea. Babel is everywhere, from protocols, connectors, and programming languages to product variants, and user interfaces. People just love variation and customization.

Let's wrap up this chapter with a brief low-level look at digital signals. Just like analogue signals, factors such as wiring and input/output impedances determine the reliability and maximum bandwidth of a communication link. High-speed links operate at frequencies comparable to radio signals and must be treated accordingly.

However, beyond their analogue characteristics, digital signals must also meet strict timing requirements. When a device detects a data request, it must respond within a specific time delay. During this period, it must fetch the data, place it on the wire(s), and hold it long enough for the receiving device to read it. A data request might be triggered by a signal transitioning high or low. Furthermore, chances are that the requested data consists of more than one bit. This raises a critical question: which bit should be sent first? As you can see, there are many opportunities for errors and mistakes to creep into digital communication.

The Printed Circuit Board & Wiring

When asked to describe an electronic circuit board, most people will probably picture some sort of card or board with strange-looking components mounted on it. This is indeed what electronics looks like today: a board—usually green—of varying shapes and dimensions with components soldered to it. But it hasn't always been like this. In the early days of electronics, circuits were relatively simple and comprised only a few components. Interconnecting them was easily done by either soldering them directly together or by soldering wires between them.

When a design was more complex, such as a TV set, often a sheet metal frame was used. On the top side, large components like vacuum tubes, transformers, and big capacitors were mounted, while the underside housed smaller resistors, capacitors, and wires. The tubes were placed in sockets, and the smaller components and connecting wires were soldered to soldering lugs or terminal strips, with wires running to the tube sockets and the terminals of other large components. Manufacturing products this way was labour intensive and prone to errors, yet it remained common practice until the 1950s or so.

However, already in the beginning of the 20th century people experimented with laminating or plating insulating supports with metal conductors either by gluing, spray-painting or by chemical processes like electroplating to simplify wiring. Some even went as far as printing components like resistors, capacitors and inductors.

In 1936, Austrian inventor Paul Eisler (1907–1992) built a two-valve radio that used wires printed as metal strips on a sheet of Bakelite. Living in England, Eisler presented his invention to Plessey but was turned down. At the time, the wiring work at Plessey was done by women and "girls were cheaper and more flexible." Eisler, however, did not give up. During the early years of World War II, Eisler developed the process of etching a copper-laminated slab of insulating material, with the circuit wiring printed on it using etch-resistant ink[34]. The modern printed circuit board, or PCB as it is commonly called, was born. Eisler's foil technique for prefabricating wiring only required standard printing equipment. Although demonstrated to the British Army, it was the American Army that recognized the PCB's potential and began using it in the production of weapons. After the war, it spread into the civilian in-

34 Paul Eisler is also credited with the invention of the rear windscreen heater, and a pizza warmer to keep take-away pizza warm while driving home.

dustry where it revolutionized electronics manufacturing.

To create a PCB using the foil technique, an etch-resistant film representing the circuit wiring must be produced. Several methods have been developed for this, some more practical than others, but they all shared one commonality: the circuits were drawn by hand. Additionally, to allow for detailed designs, they were first drawn at a larger scale and then reduced to actual size. This process made it easy for mistakes to creep in, especially as more processing steps were added. This is where Gerber enters the arena.

Heinz Joseph Gerber (1924–1996) was an American inventor and businessman who made it big in the business of drafting and cutting of various materials. He and the company he founded revolutionized the mass production of, amongst others, shoes and clothing by developing machines and techniques that could automatically and reproducibly draft and cut leather and fabric, but also aluminium for building planes and ships. Gerber was a pioneer in the field of computer-assisted manufacturing (CAM). When, in 1964, Gerber replaced the print and cut head of his drafting machine by a light beam, he created the photoplotter, the world's most precise printer. It allows transferring drawings like PCB designs directly onto photosensitive film with a resolution in the micrometre range and at high speeds. PCB manufacturing had suddenly become much faster, much easier, and much more precise. The image file format initially used for the photoplotter, based on the now obsolete EIA RS-274-D standard, became known as the Gerber format.

Traditionally, electronic components had leads and every lead had to pass through a hole in the PCB to be soldered to the board on the opposite side. This method is known as through-hole technology (THT). However, drilling holes is a time-consuming process, and the automatic placement of THT components on a circuit board is slow and complicated. Additionally, the leads make components bulky, hindering efforts toward miniaturization. Starting in the 1960s, new component packages were developed that eliminated leads, replacing them with small tabs or contact pads that could be soldered directly to the board. By the late 1990s, surface-mount devices (SMDs) had become the dominant technology, enabling ultra-fast and precise circuit board assembly. To give you an idea, current (2025) pick-and-place machines can place up to 100,000 components per hour.

To keep up with evolving technology, and in particular with the emerging surface-mount technology, the Gerber format was extended to RS-274X in 1998. Extended Gerber is currently the industry standard for describing images for PCB manufacturing, even though other formats exist, like ODB++.

Complex, high-speed, and high-power circuits all require PCBs capable of meeting their specific signal routing requirements. As a result, the printed circuit board is no longer merely a method for prefabricating circuit wiring; it has become an integral part of the circuit and circuit design itself. A PCB can make or break a product. When designed properly, it can ensure a product passes electromagnetic compatibility (EMC) testing, achieves maximum data transfer speeds, and carries high currents without risk of overheating or failure. If designed badly, a product may simply fail to work.

The Gerber file format lists only lines, traces, and pads; it does not contain information about the circuit itself or the mechanical aspects of the design. As a result, this additional information must be communicated separately from the circuit designer to the PCB manufacturer, increasing the risk of mistakes and errors.

PCB Art

Printed circuit boards have always inspired artists, as illustrated by this magazine cover from 1974. PCB art is a fusion of electronics and design, where the traces, silkscreen, and components are arranged to form artistic patterns or functional artistic pieces. Tutorials explaining how to create artistic PCB designs are plentiful on the internet, and PCB manufacturers are increasingly offering services for artistic expression, such as multicolour solder masks and full-colour silkscreen printing. Note that PCB production files are usually referred to as artwork.

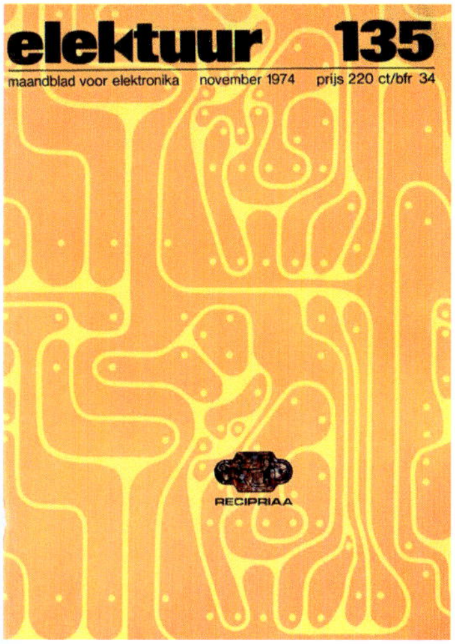

Circuit design tools provide numerous error-checking features to ensure that the transformation of a schematic into a PCB is as accurate as possible. Heat distribution and EMC radiation patterns can be simulated, and missing connections or clearance violations are flagged. However, if the PCB manufacturer is not informed that the board requires extra-thick copper laminate, the entire production batch could end up being scrapped.

Incorporating more information about a design into the production files can help avoiding silly mistakes, which is why other, more complete, PCB design file formats like ODB++ exist. However, they struggle to gain traction. The Gerber format was extended to X2plus to address this situation. In 2020, version X3 was released that adds component placement data.

With the optimization of PCB manufacturing techniques, the cost of producing a board has dropped dramatically. Originally, each customer's order was processed individually, and high volumes were required to achieve a reasonable price per board. However, in the 1990s, PCB pooling emerged as a production method. This approach combines board designs from different customers into a single manufacturing panel. PCB pooling allows production setup costs to be shared among multiple customers and optimizes the use of available panel space. As a result, smaller production runs became possible at a fraction of the cost of traditional processes. Today, PCB manufacturing is so affordable that the main cost factor of producing a few boards for prototyping is determined by the shipping expenses.

The production, use, and disposal of printed circuit boards have significant environmental impacts, primarily due to the materials, chemicals, and processes involved. PCBs contribute significantly to electronic waste (e-waste), one of the fastest-growing waste streams globally. Environmental standards like RoHS (Restriction of Hazardous Substances)[35] and WEEE (Waste Electrical and Electronic Equipment Directive) attempt to mitigate the risks.

The most commonly used material for PCB substrates is known as FR4, a fiberglass-reinforced epoxy laminate. FR4 and other PCB materials are not biodegradable, persisting in landfills for centuries. Burning PCBs to reduce waste volume releases harmful chemicals, such as brominated dioxins and furans. Electronic assemblies contain important quantities of metals like copper and tin, but also gold, silver, mercury and cadmium, to name a few, and improper disposal can lead to soil and water contamination.

FR4 is still the go-to material for PCBs because of its reliability and affordability, but there are greener alternatives like natural fibre composites, recycled materials, and ceramic-based options. These eco-friendlier choices are becoming more attractive as the electronics industry looks for ways

35 RoHS limits the use of lead to a maximum concentration of 0.1% by weight in any single material, including solder. As a result, most electronic products are required to use lead-free solder.

Why are PCBs Green?

The colour of printed circuit boards (PCBs) is determined by the solder mask, a protective layer applied over the copper traces to insulate them and protect against oxidation, contamination, and accidental solder bridging. The typical green colour associated with PCBs originates from the early epoxy resins used in solder masks, which were naturally green. By fortunate coincidence, the human eye is particularly sensitive to green. As green provides excellent contrast against the white silkscreen markings and the shiny copper traces, green PCBs are more suitable for manual inspection and troubleshooting. Over time, the green solder mask has been refined and has become the most cost-effective and reliable option.

Today, PCBs can be manufactured in a variety of colours, including red, blue, black, white, purple, and yellow. Choosing a colour other than green can not only improve perception and branding but may also influence certain properties of the board. For example, a black solder mask can enhance thermal conductivity and UV resistance.

The silkscreen is typically white, but full-colour silkscreens are also available. This allows the reproduction of, for instance, photographs on a circuit board, overriding the colour of the solder mask.

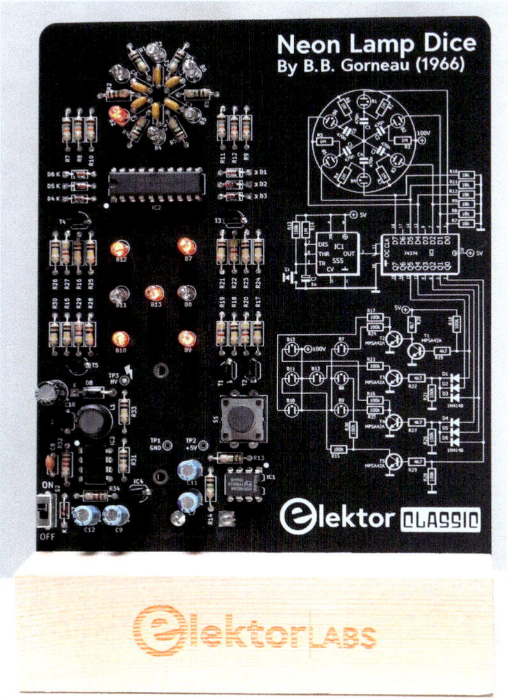

A PCB-based decorative object.

to lessen its environmental impact. That said, factors like higher costs, limited scalability, and performance trade-offs are still hurdles to their broader use Printed circuit boards (PCBs) provide the electrical connections between the components of a circuit. However, wiring does not stop there. In many cases, a circuit board must be connected to external components that cannot be mounted directly onto it. These may include user interface elements such as buttons, switches, and displays, or embedded components like sensors and relays positioned deep inside a machine. Connections between multiple boards and modules are also common. When a circuit is too large to fit on a single board, multiple interconnected boards are required. Another frequent scenario is when two or more boards must exchange data or share power.

The method of electrically connecting two devices depends heavily on the type of signals being transmitted. Sometimes, a simple wire suffices, but in most cases, a more sophisticated interconnection system is necessary. Selecting the appropriate wires, cables, and connectors is crucial to ensuring reliability and performance.

To appreciate the diversity of interconnection systems, consider power distribution networks and motor drivers, high-speed digital communication interfaces like USB and HDMI, low-power analogue sensor signals, miniature connectors in smartphones, vibration-resistant wiring in automotive and maritime applications, and audio-video transmission. These all fall under the domain of the interconnect and electromechanical components industry

Besides visible wiring as printed on PCBs and running between subsystems, there exists a third type of wiring: inside integrated circuits. Not only he transistors that make up the circuit must be connected together, also the chip's in- and outputs must be bonded to the ICs contacts to the outside world. Doing this is not an easy task, especially when there are hundreds of them.

Interconnecting the transistors on a chip may seem trivial at first glance as the they can simply be etched or deposited onto the silicon substrate, similar to a PCB design. For simple circuits this is true, but what about complex high-density digital systems like microcontrollers, microprocessors and FPGAs with hundreds of contacts and billions of transistors? How do you connect 100 million gates or more of an FPGA in such a way that you can reconfigure it any way you want? How do you ensure that every bit inside a multimillion-bit memory array can be accessed at the same speed? How do you distribute power? How to respect the timing requirements of superhigh-speed clock signals?

Barrel Jack Hell

One of my pet peeves (and I have many) is the barrel jack connector, also known as the DC power jack. Commonly used by plug-in AC-DC power adapters (a.k.a. wall warts) to provide power to low-voltage devices such as radios and toys, it consists of a barrel (sleeve) with a centre pin on the jack and a corresponding hole in the plug. While similar in design to coaxial and RCA (cinch) connectors, it is not identical. Determining exactly when barrel jacks first appeared is difficult, but they seem to have been introduced as early as the 1950s, 1960s, coinciding with the rise of portable, battery-powered electronic devices.

The barrel jack has at least two major problems. First, there are too many connector variations. At the time of writing, Wikipedia lists 60 different barrel jack sizes. While size does matter for voltage and current handling, a standardized set of three to five sizes would have been sufficient for most use cases.

The second problem is voltage polarity confusion. Today, the most common configuration is centre-positive (pin = +, barrel = –), but centre-negative also exists (of course 🙄). And this is my pet peeve. To me, the centre pin should always be negative (or ground), as it is the first point of contact when a jack is inserted into a socket.

Furthermore, many barrel jack sockets include a built-in switch, typically used to disconnect a device's internal battery when an external power source is plugged in. This switch is wired to the barrel, not the pin. If the barrel is connected to ground, this design momentarily disconnects the ground connection when inserting the plug—something that should never happen in a well-designed circuit.

I'm sure designers have all sorts of justifications for making the centre pin positive, but that doesn't make it any less wrong.[36]

36 The first point of contact of coaxial and RCA connectors is the shield. True, this is not the centre pin, but when examined closely, you will notice that the pin makes contact after the shield.

The popular 4060 14-stage binary counter and oscillator has ten outputs, ranging from Q3 up to Q13. At first glance, this seems to total eleven outputs, but Q10 is missing. While it exists internally, it is not connected to an external pin, likely due to constraints in fitting the design into a 16-pin package. However, why Q10 specifically was omitted, rather than another output, remains a mystery.

Wiring a chip can be a very complicated matter indeed. In current high-density chips, the wiring alone can consume up to 80% of the chip's real estate. While transistors keep shrinking in size, scaling down wires is difficult as this introduces impedance and power capacity issues. Thinner wires have higher resistance and capacitance, increasing signal propagation delays, voltage drop and heat production. Furthermore, high-capacitance coupling between adjacent wires leads to signal integrity issues (crosstalk noise). It's like a crowded metropole, but worse. To improve things, materials to replace copper are being researched and different chip architectures are tried.

An obvious technique for wiring up a circuit is by connecting wires between the components that make up the circuit. Discrete wiring has been done ever since the beginning of (electronics) time. Instead of soldering wires to components, it is also possible to wrap wires around the component leads. The resulting technique is known as wire wrap. Technology being technology, wire wrap has evolved to a point where it can produce more reliable results than those obtained with printed circuit boards. Wire wrap is still being used today for high-reliability circuits in aerospace and defense. It is also practical in rapid prototyping as it allows for quickly wiring complex circuits that can be modified easily.

In wire wrap, components use long square leads called (terminal) wrapposts on which more than one wire can be wrapped. Tightly wrapping a solid wire around a square post creates strong contacts due to the post corners biting into the wire (cold welding). A NASA-compliant Class-A wire wrap contact consists of one half to one-and-a-half turns of insulated wire in contact with a minimum of three corners of the post, followed by typically seven turns of stripped wire. This creates 28 contact points. The exact number of turns depends on the wire diameter. The wrapposts must be gold or tin plated.

Wire wrapping can be done quickly and precisely by machines, but simple manual tools can produce great results too.

The Internet

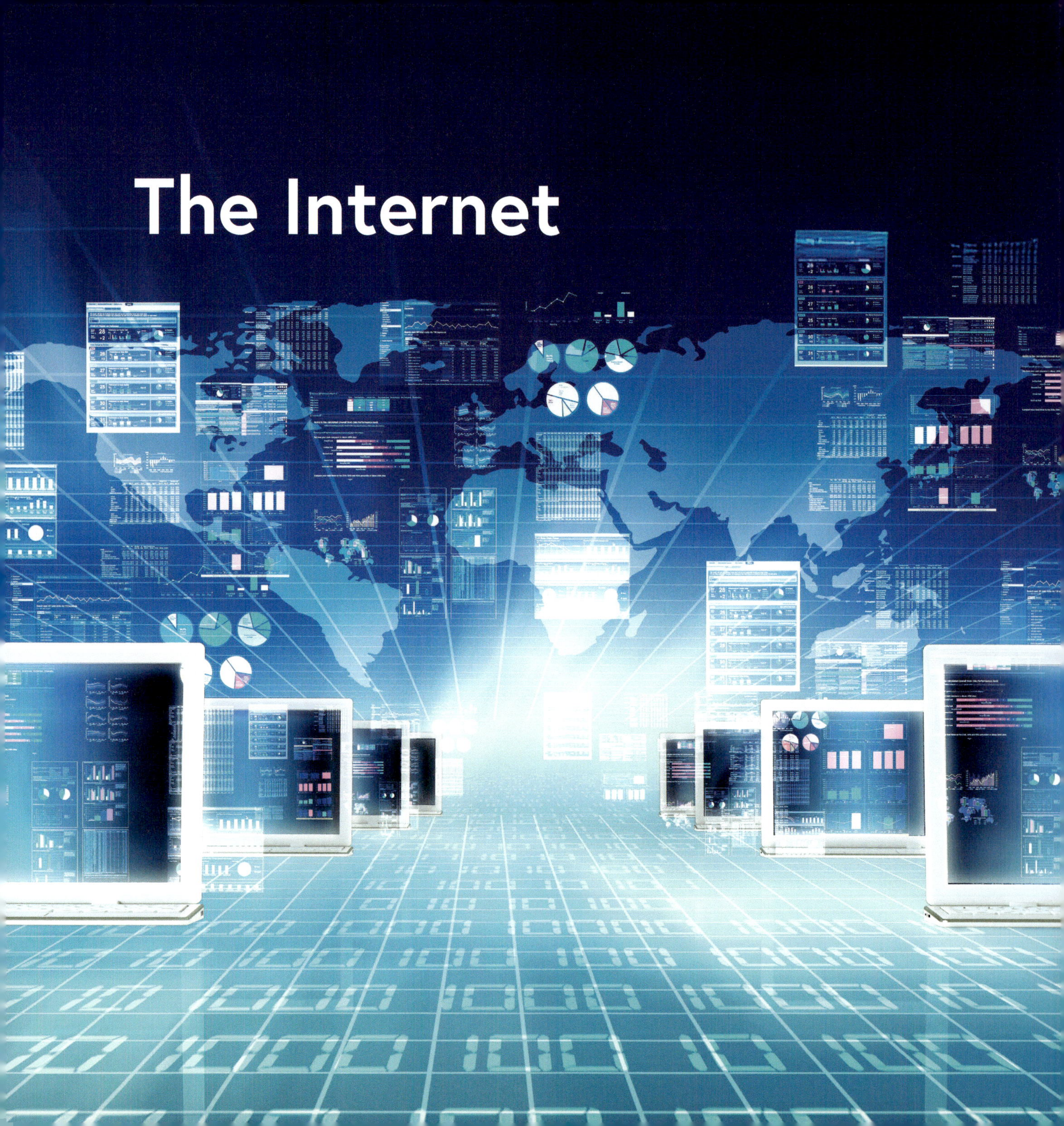

You may find it surprising to see a chapter discussing the internet in this book. While it's true that the internet as we know it wouldn't exist without electronics, that isn't the focus of this chapter. The reason for including it is that the internet has become an essential tool for the electronics industry. This is, of course, true for almost everything today, but that doesn't make it any less relevant to this book.

Indeed, the internet has revolutionized the way we work in nearly every field, and electronics is no exception.

The internet was not invented by a single person but is the result of contributions from multiple scientists, engineers, and organizations, beginning in the 1960s and continuing over several decades. The prototype of the internet was ARPANET, the first operational network to use packet switching. In 1983, the TCP/IP protocol, developed by American engineers Robert Elliot Kahn (1938–) and Vinton Gray Cerf (1943–), was adopted, marking the official birth of the internet. Later, English computer scientist Sir Timothy John Berners-Lee (1955–) invented the World Wide Web (WWW), including HTML, HTTP, and the first web browser, making the internet more accessible and user-friendly.

Mother of the Internet

Many individuals played crucial roles in shaping the internet as we know it today. Those who made significant contributions have been honoured with the title of Internet pioneer. Among them is American computer programmer and network engineer Radia Joy Perlman (1951–) who invented the spanning tree protocol (STP), which is fundamental to network bridges and Ethernet networks. This achievement has earned her the (unofficial) title of "Mother of the Internet."

Before the internet, finding information was complicated and time-consuming. Engineers had to rely on datasheet books, which contained hundreds of thin pages filled with essential specifications, graphs, and application examples. These books were handed out sparingly by component suppliers, typically only to potential high-volume clients. Even harder to obtain were application notes, which provided detailed circuit descriptions. For non-professional electronics enthusiasts, such documentation was simply *unobtainium*. Today, thanks to the internet, this problem has vanished, along with the datasheet books. Product documentation tends to be

plentiful and readily available for download directly from manufacturers' websites.

Not only has the internet made access to product and application documentation easy, but it has also sparked a commercial revolution. In the past, a selection of popular components could only be purchased in small quantities from specialized shops. Today, almost anything can be found online in any quantity. Thanks to the internet, it has become possible to request free product samples directly from a component manufacturer instead of begging for one from a distributor.

With access to both documentation and physical parts becoming significantly easier, a new market has emerged where obscure manufacturers sell small, unbranded preassembled modules designed for integration into hobby projects and prototypes. Thousands of people share their projects based on these modules online, enabling others to learn and start tinkering for themselves.

Up until the 1990s, amateur electronics was booming. However, with the rise of personal computers, interest in electronics began to wane as people shifted their focus to programming and software development. As a result, most electronics magazines and specialized shops disappeared. Then came the internet, and a few years later, the maker emerged. More than just an electronics hobbyist, a maker is someone who creates using a blend of traditional craftsmanship and modern technology. Alongside this, a distinct maker culture developed, a movement that values hacking existing devices and sharing software as open source. Open hardware also became a part of this culture, typically consisting of freely available designs for electronic devices that can be copied, modified, and sometimes even commercialized.

Starting out small, the maker movement has gained such momentum that the electronics industry could no longer dismiss it

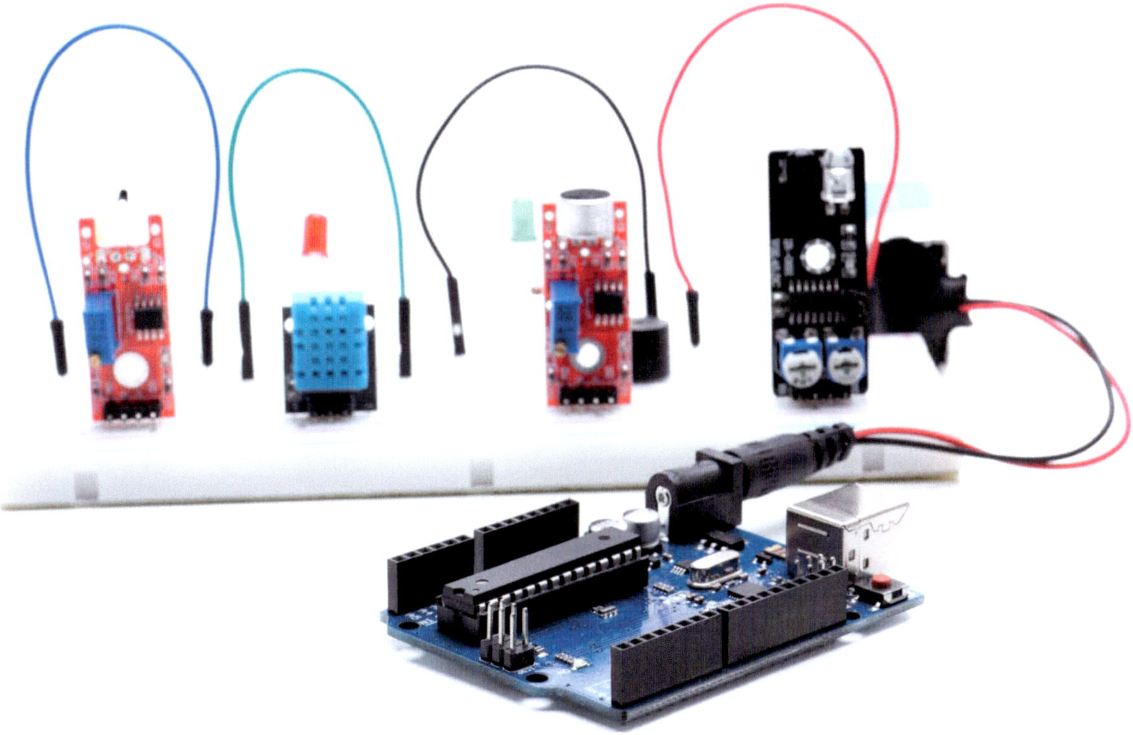

as it once did with hobbyists. Today, semiconductor manufacturers actively engage with the maker movement to discover new application ideas and promote emerging components. In hindsight, it has become clear that companies like Apple and Intel were founded by makers. Many of the individuals who made significant contributions to the electronics industry were makers, and many professional engineers continue to be makers today.

The internet and the open-source movement have also led to the availability of cheaper and more advanced tools for designing electronic circuits. Schematic capture and PCB design software are now available for free, as are circuit simulators. PCB pooling services would not have been possible without the internet. Likewise, software toolchains for developing embedded applications—which once cost thousands of euros—can now be downloaded for free. As a result, amateur

A product of the open-source movement, the Arduino development platform created in 2005 made microcontrollers accessible to people without an engineering background. Nearly twenty years later, it has not only been embraced by millions of makers worldwide but has also been adopted by many semiconductor manufacturers to promote their own products

The Capacitor Plague

Between 1999 and 2007, an unexpectedly high failure rate of non-solid electrolytic capacitors was observed in computer motherboards, video cards, and power supplies. In particular, capacitors from certain Taiwanese manufacturers proved to be highly unreliable. The root cause was traced to faulty electrolyte, which did not have the correct chemical composition.

This led to corrosion, generating gas that caused the capacitors to bulge, leak, or even explode.

According to some reports, the so-called capacitor plague was the result of industrial espionage. Allegedly, a capacitor scientist who moved from Japan to China took with him the formula for a high-performance electrolyte. His new Chinese colleagues attempted to replicate the formula but, unknowingly, made errors in the process. Later, some of these scientists defected to Taiwan where they sold their flawed electrolyte formula to several capacitor manufacturers who used it in their products. As a result, they failed prematurely, affecting countless electronic devices. To this day, the exact number of affected products remains unknown.

electronics is thriving like never before. At no point in history have so many people been actively engaged in electronics as there are today[37].

Of course, the internet and globalization also have their downsides. For the electronics industry, one of the most persistent challenges for decades has been the grow-

37 Percentagewise, of course, as the world population continues to increase and with it the number of electronics engineers.

ing problem of counterfeit components. As the electronics industry expanded, so did the demand for cheaper parts, creating opportunities for counterfeiters to infiltrate the supply chain.

Counterfeit components first appeared in the 1970s and 1980s, when electronics manufacturing began shifting to outsourced production in Asia, particularly in China and Taiwan. Early counterfeits were often used components that had been rebranded to appear new. Today, recycling components from e-waste and selling them as new has become common practice.[38]

Another form of counterfeiting involves producing low-quality clones of name-brand components. While counterfeiters initially focused on expensive parts, even low-cost components like capacitors and transistors are now frequently cloned and faked. Fake chips today include refurbished chips with fake date codes, fake microcontrollers with non-functional firmware and components with internal defects or poor-quality silicon.

As counterfeiting techniques become more sophisticated, identifying and avoiding fake components is becoming more and more difficult.

A counterfeit 4580 low-noise amplifier IC? The part on the right is genuine.

38 But isn't that actually a good initiative that should be praised and followed?

A Look into the Future

As is evident from reading this book, the development of the electronics industry has been driven primarily by humanity's urge for communication. Many, if not most, of the technologies, materials, tools, and gadgets we enjoy today are the result of research in telecommunication. Some, like optical fibre, are directly used in communication, while others, such as LED lighting, are byproducts. The list of examples is virtually endless. Therefore, it is reasonable to assume that this trend will continue in the future. However, predicting, let alone accurately guessing, the state of technology in the coming years or centuries is impossible. Short-term predictions tend to be overly optimistic, whereas long-term forecasts are often too conservative.

By 2189, 164 years from now, the hottest smartphone of 2025 might be as obsolete as the 164-year old Pantelegraph fax machine from 1861 is today. The art of predicting the future lies in thinking outside the box and disregarding existing technology. Imagine a communication device that is nothing like a smartphone—something you don't need to carry around and that can't accidentally fall into the toilet. What capabilities would your ideal communication device or technology have?

The Pantelegraph invented in 1861 by Italian priest Giovanni Caselli (1815–1891) was the first fax machine. Interestingly, faxing preceded the transmission of speech. Scottish inventor and engineer Alexander Bain (1810–1877) too worked on an experimental fax machine as early as 1843. He was also the first to patent the electric clock (yet another example of a patent scandal).

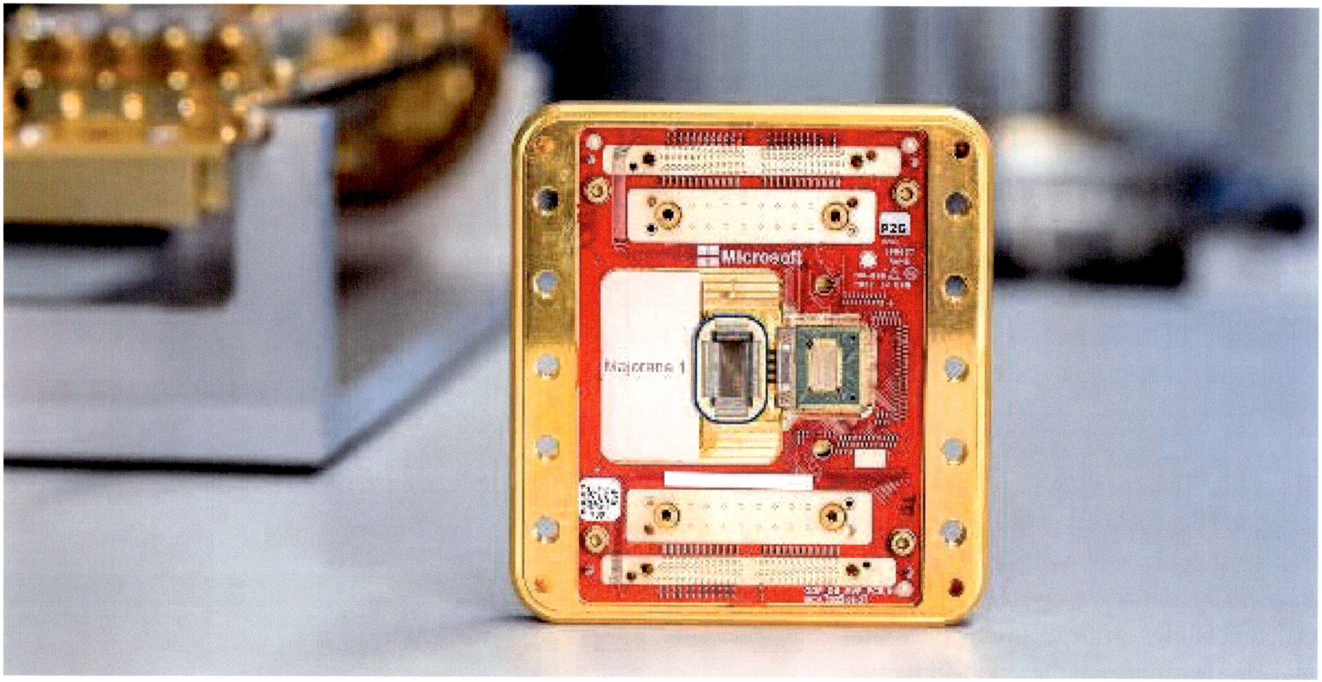

The Majorana 1 quantum chip by Microsoft is based on topoconductor or topological superconductor, a special material which can observe and control Majorana particles, a particle that does not exist in nature. The material allows producing more reliable and scalable qubits. The device has eight topological qubits and allows scaling up to one million.

mechanical properties make it a promising material for transistors, sensors, and flexible electronics. Similarly, the development of new materials like GaN (Gallium Nitride) and SiC (Silicon Carbide) have led to more energy-efficient devices, machines, and electric vehicles.

When a sheet of graphene is rolled up, it forms a cylindrical molecule known as a carbon nanotube (CNT), which possesses exceptional electrical, thermal, and mechanical properties. As the name suggests, CNTs are extremely small, enabling ultra-miniaturised circuits. Furthermore, as they conduct electricity more efficiently than copper, CNT-based transistors could potentially outperform silicon transistors in both speed and power efficiency. CNT FETs were first demonstrated in 1998, but a suitable technology for their mass production has yet to be developed.

A step beyond CNTs is molecular electronics which uses single molecules as electronic components, such as transistors, diodes, and switches. Instead of silicon, these devices rely on the electronic properties of individual molecules. Molecular electronics enables circuits that are far smaller than those achievable with silicon technology. It has the poten-

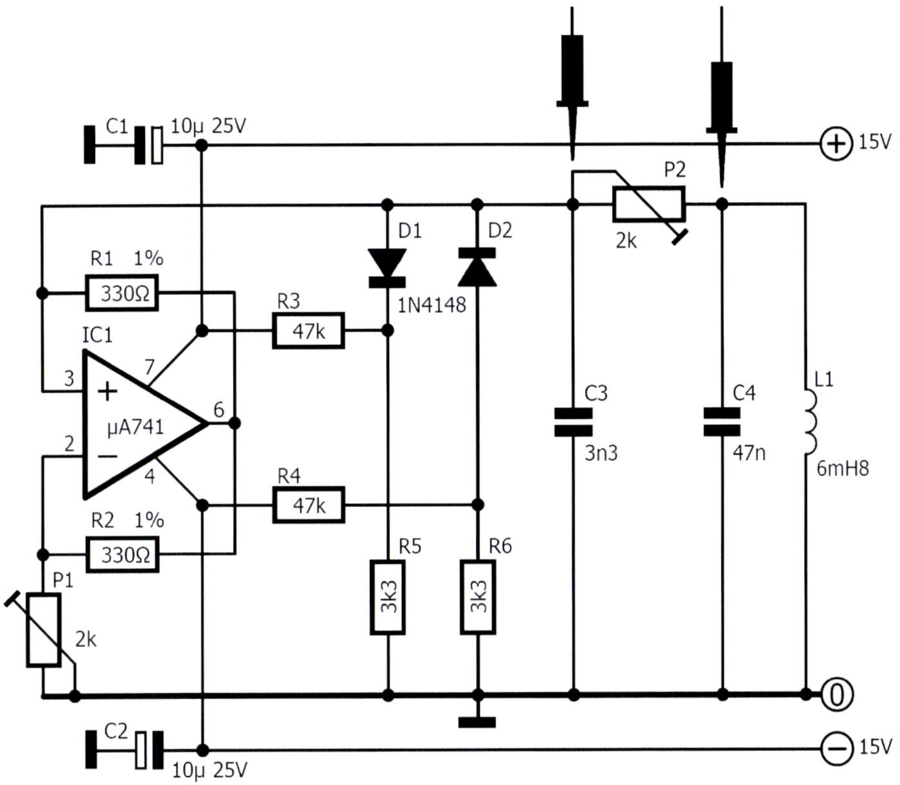

In 1983, Chua invented a non-periodic oscillator that became known as Chua's circuit. It is the simplest electronic circuit that exhibits chaotic behaviour. One of its applications is in chaos computing where it is used to create morphable logic gates with. A morphable gate can be reconfigured dynamically to obtain any type of logic gate. Systems based on chaotic logic gates can reconfigure almost instantly and enable reconfigurable hardware. Different circuits for morphable gates have been proposed, the most famous being the ChaoGate.

tial to create computers at the molecular scale, operating with significantly lower energy consumption than traditional semiconductors.

Besides new materials for semiconductors, it is also possible to invent new components that may transform the landscape of electronics. In 1971, Filipino engineer and computer scientist Leon Ong Chua (1936–) postulated the memristor, the fourth fundamental electrical component, alongside the resistor, capacitor, and inductor. The resistor relates voltage and current, the capacitor relates voltage and charge, while the inductor relates magnetic flux linkage[40] and current. The memristor, on the other hand, relates magnetic flux linkage and charge. A memristor's resistance changes depending on the amount and direction of charge that has passed through

40 Flux linkage refers to the interaction of a multi-turn inductor with the magnetic flux as described by Faraday's law of induction.

it. If power is removed, the memristor remembers its last resistance state, making it a non-volatile memory device. Unlike traditional binary memory, memristors can store multiple resistance states. Not only does this enable higher-density data storage than conventional memory technologies, but it also makes them suitable for analogue computing and neuromorphic applications.

However, the memristor is also a controversial device. Some scientists believe that the memristor is merely a theoretical concept, whereas Chua argues that memristor-like behaviour is common and had already been observed by Sir Humphry Davy (1778–1829) in the early 1800s, even before the resistor and inductor were formally recognised. Fact is that memristor-like devices do exist and can be bought and used in circuits[41].

Related to memristors is Resistive Random-Access Memory (ReRAM), a type of non-volatile random-access memory that operates by changing the resistance across a dielectric solid-state material. Although it was once predicted to have a bright future—in 2013, Hewlett-Packard forecast that 1.5 PB (one quadrillion bytes) ReRAM solid-state disks (SSDs) would be available by 2020—ReRAM adoption has not progressed as quickly as hoped.

Similar to how the memristor completes the theoretical quartet of fundamental electrical components, the trancitor completes the theoretical quartet of elementary active devices. Currently, only two are known: the bipolar junction transistor (BJT) and the field-effect transistor (FET). The BJT is a current-controlled current source (CCCS), while the FET is a voltage-controlled current source (VCCS). The trancitor, a contraction of 'transfer-capacitor', would function as either a current- or voltage-controlled voltage source (CCVS or VCVS). The trancitor, proposed in 2018 by South Korean engineer Sungsik Lee (1976–), has the potential to simplify transistor circuits, thereby extending the validity of Moore's Law. As far as I know, the trancitor currently exists only as a theoretical concept, but perhaps one day it will simplify our engineering lives.

Molecular electronics, as mentioned above, steers electronics away from semiconductors, but one could also imagine replacing electrons with something else. Silicon photonics, for instance, uses photons (i.e., light) instead of electrons to transmit and process data within silicon-based circuits.

41 See https://knowm.org/

Since light travels faster than electrons, photonic circuits offer higher speeds and greater bandwidths than traditional electronic circuits. Silicon photonics is also more energy-efficient and generates less heat. Moreover, silicon photonic devices can be manufactured using existing semiconductor production techniques. Hybrid devices integrating optical and electronic components in a single microchip are therefore relatively easy to build.

Instead of replacing electrons with another particle, an alternative approach could be to use electrons themselves in a different way. This is the basis of spintronics, a technology that utilises not only the charge of electrons but also their spin to store, process, and transfer information. Spintronics consumes less energy than charge-based electronics, and data remains stored even when power is turned off. It also has connections to quantum computing, as qubits can be created using electron spin. In 1990, Indian scientists Supriyo Datta (1954–) and Biswajit Das proposed the spin field-effect transistor (SpinFET). Inspired by photonics, the spin transistor uses magnetic contacts to manipulate the spin of electrons, where photonics plays with the polarisation of photons.

Electron spin and spintronics are closely linked to magnetism. This has led to the emergence of magnonics, a field that studies and manipulates magnons instead of electrons or photons. Spintronics still relies on electrical currents, magnonics does not. A magnon is a quantised spin wave, which is a collective excitation of electron spins in a magnetic material. Instead of moving electrons, magnonic devices use waves of spin fluctuations to transfer information. This eliminates resistive losses, making magnonics an energy-efficient alternative to traditional electronics that enables ultra-low-power, high-speed, and nanoscale computing.

Although it is probably possible to extend this list with other obscure nanotechnologies with fantastical names, I'll stop here. Will any of the advanced technologies presented in this chapter one day become as significant as electronics, or will they be forgotten in a few centuries? Have I overlooked something important? Only the future will tell—but I doubt I'll still be around to hear the answer.

If I say electrons behave like particles I give the wrong impression; also if I say they behave like waves. They behave in their own inimitable way, which technically could be called a quantum mechanical way. They behave in a way that is like nothing that you have seen before.

Richard P. Feynman

Electronics Engineering is Also...

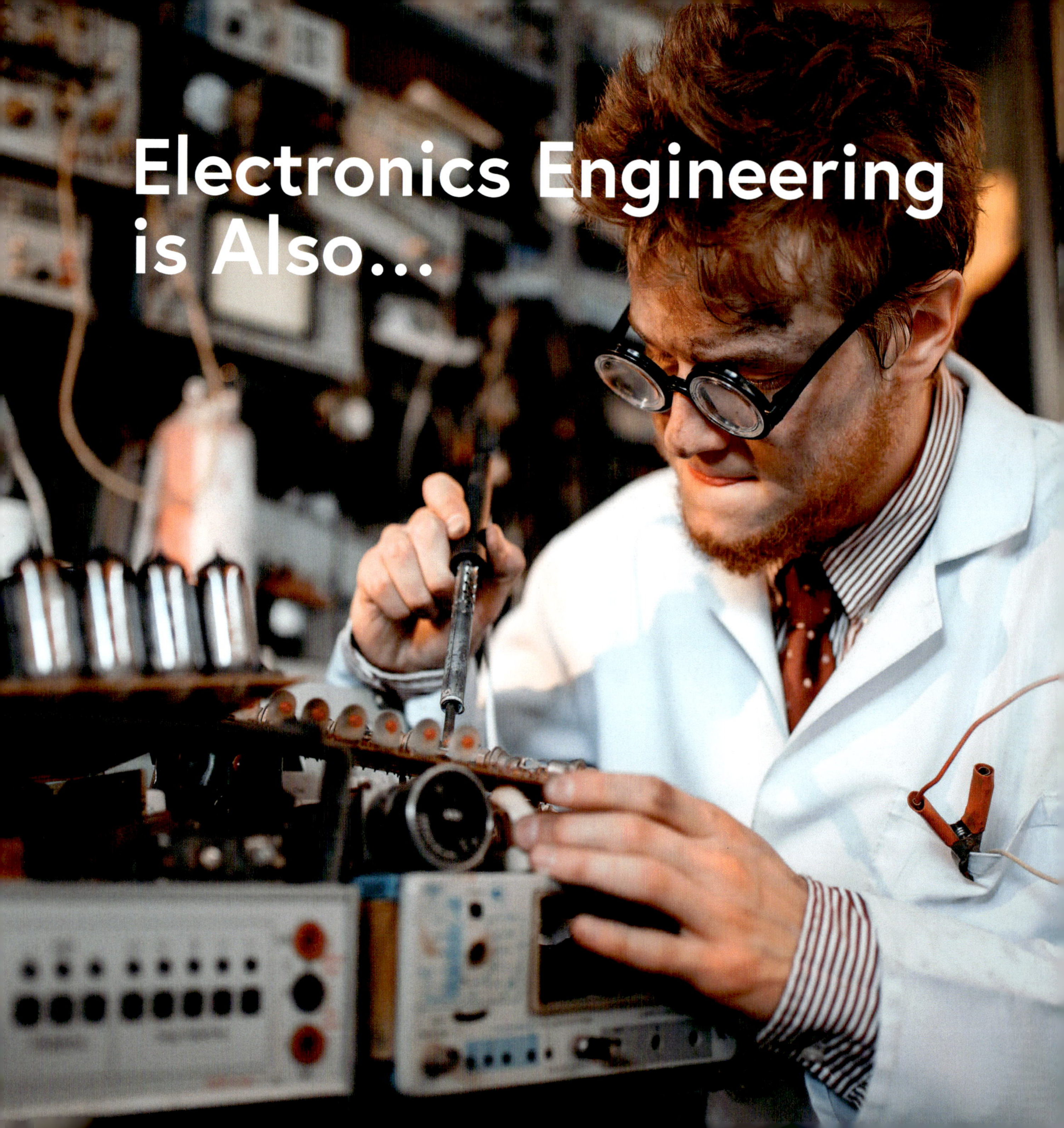

The five senses of electronics engineering. If they don't kill you, they will make you stronger.

List of Illustrations

72 - Texture wave oscilloscope by ALEXEY FILATOV (Adobe Stock 109017046)
73 - Miner give up and fed up after almost digging and reaching to the gold or treasure in the underground tunnel. by leremy (Adobe Stock 607202405)
75 - Amplifier with negative feedback by the author
76 - Wireless receiving system by US patent 1113149
79 - Baron Munchausen by Zentralbibliothek Solothurn
80 - Electronic components on a circuit drawing. by Hseyin (Adobe Stock 1018789511)
81 - Op-amp symbol by the author
82 - 2-W Operational Amplifier (early version) by Jim Williams, http://www.philbrickarchive.org/k2-w_operational_amplifier.htm
83 - Discreet operational amplifier by Elektuur, 1966
85 - The op-amp as inverting amplifier by the author
86 - Blue light-emitting diode by shima-risu (Adobe Stock 15077757)
87 - Blue energy beam by Yurok Aleksandrovich (Adobe Stock 2451656)
88 - Movement of microparticles by beams of laser in lab by Pavel Losevsky (Adobe Stock 45791480)
89 - Optocoupler circuit by the author
90 - The 7-segment display by US patent 974943)
92 - led lamps by kreatorex (Adobe Stock 49904352)
93 - Blue LED Character Display by NNL_STUDIO (Adobe Stock 67784100)
94 - Modern abstract black glitch tv wallpaper. by RDVector (Adobe Stock 295154229)
95 - General communication system by the author
98 - Vector scientific or educational scheme or diagram of the analog signal and digital signal isolated on white. by petrroudny (Adobe Stock 416897583)
99 - The Minivac 601, a computer trainer designed by Claude Shannon. by Autopilot (Wikimedia)
100 - Digital logical nor gate icon background with binary code background by Photoenthusiast82 (Adobe Stock 510250429)
102 - Logic gates by the author
103 - XOR gate by the author
103 - Two-way switch by the author
103 - 2D AND, OR and XOR function plots by the author
104 - 2-bit binary adder by the author
105 - Seamless pattern of transistor multivibrator... by aquatarkus (Adobe Stock 427534099)
106 - PCB with logic ICs by https://electronics.stackexchange.com/questions/326189/pcb-designof-the-80s-arcade-pcbs-groundpower-tracks and reworked by the author
107 - Majority gate by the author
108 - Microprocessor on printed circuit board by asb63 (Adobe Stock 722841448)
110 - Z80 CPU by Digipic (Adobe Stock 14390207)
111 - SDR FM radio by the author
113 - 8051 block diagram by the author
114 - Chip on board with blue backlight by asb63 (Adobe Stock 717453332)
116 - The Tower of Babel (1563) by Pieter Bruegel the Elder (1525/30-1569). by Adam Ján Figel' (Adobe Stock 327969915)
118 - Morse code by US patent 1647
119 - Peripherals VGA. Port for connecting monitor by shandor_gor (Adobe Stock 575056261)
121 - Man worships woman. by Alexandr Sidorov (Adobe Stock 284976783)
122 - 120W Fast Charging 4 Ports Mobile Phone Charger QC3.0 USB Type C by Родион Бондаренко (Adobe Stock 1244521650)
123 - Timing diagram by Analog Devices, datasheet DS28E18
124 - A printed circuit board in green by Remigiusz Góra (Adobe Stock 517336743)
127 - PCB art by Elektuur, 1974
129 - Neon Lamp Dice by the author
131 - Panic people wire adapter cables pop art background by studiostoks (Adobe Stock 179434210)
132 - 4060s by the author
133 - Macro with focus on a column of wire wrap junctions on a prototype electronic circuit board. by Kim (Adobe Stock 324762159)
134 - Internet Computer by Nmedia (Adobe Stock 64180529)
136 - Makerspace banner - STEAM Education by Angela (Adobe Stock 221144134)
137 - Four sensor for arduino by Jarp (Adobe Stock 239075256)
138 - Damaged cracked electrolytic bloated exploded capacitors isolated on white background. by Vlyaks (Adobe Stock 440143872)
139 - A counterfeit 4580 by the author
140 - Virtual reality glasses retro girl science fiction by studiostoks (Adobe Stock 113592134)
141 - Pantelegraph by Alessandro Nassiri for Museo nazionale scienza e tecnologia, Milano, CC BY-SA 4.0
143 - 3D illustration of The Evolution Of Storage Devices by vexworldwide (Adobe Stock 134840254)
144 - Quantum qubit in superposition state, bloch sphere... by atdigit (Adobe Stock 662444557)
146 - Majorana 1 quantum chip by photo by John Brecher for Microsoft.
147 - Chua's circuit by Elektor 01/12/1994
150 - Strange scientist soldering chips, test in lab by Nomad_Soul (Adobe Stock 291150432)
151 - The five senses of engineering by the author
- Back cover photo by Nathalie Valens

If you believe you should have been credited, or credited differently, please contact the editor.

List of People

List of Terms, Inventions and Discoveries